To Kirk Watson,

Thanks for all your good work for Austin, Texas, and our country.

Your leadership and accomplishments have been an inspiration to many.

# THE **FUTURE** OF **BUILDINGS, TRANSPORTATION** AND **POWER**

ROGER
DUNCAN

MICHAEL
E. WEBBER

For information about this title or to order other books and/or electronic media, contact the publisher:
DW Books
3005 S. Lamar Blvd. STE D109 #369
Austin, TX 78704-4785

ISBN:   978-1-7344290-0-8 (softcover)
        978-1-7344290-1-5 (eBook)
        978-1-7344290-2-2 (hardcover)
        978-1-7344290-3-9 (audio book)

Printed in the United States of America

Cover and Interior design: 1106 Design

*Roger: I dedicate this to the love of my life, Jo*

*Michael: This book is dedicated to everyone
who believes a better future is possible*

# NOTE ON THE CORONAVIRUS PANDEMIC

In the introduction, we point out the folly of predictions. Boy, did we get that one right. As this book was going to press the coronavirus pandemic struck, leveling the global economy. Surely much of the innovative technology that was scheduled to roll out in the coming year will be delayed, reduced in scope, or even canceled. It may be amusing to read some of our predictions about what changes are imminent.

Yet we believe the underlying technology trends are still in place, and when the world's economy fully recovers, the buildings, transportation, and power sectors will continue the transitions in place at the beginning of the year. The pandemic is not going to stop the transition to renewable energy and electric vehicles and sustainable building. Some of the technologies mentioned, such as robotic delivery service, may even be accelerated. However, many of the specific projects and initiatives will certainly be pushed further into the future. We have added updates as best we could at the last minute to address the issue.

# FOREWORD

We first started brainstorming about this book in the fall of 2008, meeting after work at El Mercado or other Austin venues to discuss our various ideas and swap notes. From those brainstorming sessions to publication 12 years later, we changed our minds many times about the title, table of contents, sequence of chapters, main arguments, and whether it should be a book, article, or series of public lectures. We did some variation of all three along the way. We often joked that we were taking so long to write a book about the future of energy that by the time it would be published, we would have to pitch it as a history of energy. As we wrote, our anecdotes about futuristic technologies (drones!) slowly moved forward into our sections on current capabilities. This sequence played out several times, so we had to hustle to stay ahead of the curve. That process forced us to stay alert and keep reading and learning. Some of our key concepts never changed, revealing themselves to be fundamental. We hope you agree and that you like this book.

Roger Duncan and Michael Webber
Austin, Texas, USA, Spring 2020

Roger (right) and Michael (left, face out of view) brainstorming about the book on October 11, 2011 at the Energy Institute at the University of Texas. At that time, the book's working title was *Clean Energy Principles and Priorities*. The red diagram at the bottom was converted into an article with Marianne Shivers Gonzales in the Winter 2013 edition of *Issues in Science & Technology* titled "Four Technologies and a Conundrum: The Glacial Pace of Energy Innovation."

During a brainstorming session on October 11, 2011, Roger lays out the core underlying concept of the book.

# CONTENTS

## PART 5

# INTRODUCTION

## THINKING ABOUT THE FUTURE

Most people at some time or another have fantasized about stepping into the future, if only for a moment. Anyone who grew up watching *The Jetsons* knows the feeling. Would there be flying cars? Would buildings be sleek, "smart," and clean, or would they be just one more dysfunctional component of a decaying infrastructure? Would there be robots everywhere? Would life be peaceful or conflict-ridden? *Star Trek* or *The Terminator?*

This book is our effort to describe the energy future. It presents our vision for buildings, transportation systems, and the electric grid.

These three sectors—buildings, transportation and the generation of electricity—account for more than 75 percent of the total energy consumed in the U.S., and the same is generally true worldwide.[1] Although agriculture and industry are also important energy consumers, most types of energy conversions take place in these three sectors, making them useful lenses into the future. Moreover, we believe that the increasing interconnections of energy and information in these sectors present a fascinating story about our future.

## THE ORDER OF THIS BOOK

Part 1 introduces the technology megatrends that we think will shape the buildings, transportation, and power systems of the future. This chapter also demonstrates how the increasing information intensity of our technology will result in sentient-appearing machines and the convergence of the three sectors. We also discuss some caveats on what we can expect from technology.

Part 2 focuses on the future of buildings. Chapter 2 discusses building trends—what we can expect if we keep doing things as we have been doing up until now. The chapter talks about how buildings are becoming more energy efficient and better connected. At this stage, buildings are becoming "smart." This chapter also discusses the trend of installing solar or other on-site generation to meet power needs.

In this book, we rely heavily on data from the International Energy Agency (IEA) and the Energy Information Administration (EIA). For the sake of avoiding an alphabet soup-style confusion, suffice it to say that the IEA, as its name states, is a Paris-based international, intergovernmental operation that, according to its website, "Taking an all-fuels, all-technology approach, the IEA advocates policies that enhance the reliability, affordability, and sustainability of energy."[2]

By contrast, the EIA is a division of the United States Department of Energy, tasked with collecting, analyzing, and disseminating "independent and impartial energy information to promote sound policymaking, efficient markets, and public understanding of energy and its interaction with the economy and the environment."[3]

There are slight differences in the assumptions used by the agencies in their projections, and the reader can refer to those differences in the endnotes.[4, 5] Usually, when we are talking about U.S. data, we will be using the EIA information, and global energy trends will be referenced from the IEA.

The third chapter concerns itself with sustainable building. It shows the many ways buildings can be made more energy efficient than today's buildings, as well as meeting more of our energy needs on-site with clean resources. While not yet sentient-appearing, buildings can fulfill their functions more sustainably through "intelligent efficiency."

Chapter 4 is about the rise of sentient-appearing buildings—namely, how the increased information technology in our buildings will enhance their functions, make them responsive to climate change, lead to autonomous behavior, and develop a new and evolving relationship between buildings and their occupants.

Part 3 focuses on the future of transportation. Chapter 5 is about all the different ways we are displacing oil-based transportation fuels, and the obstacles involved. This includes a sustainable transportation sector as achieved through cleaner fuels and technologies.

The sentient-appearing transportation system of the future is the subject of Chapter 6. The system will always be on, aware of where you are and where you need to go, and efficiently move people and goods around the globe—in some cases very quickly.

Part 4 examines the future of power. Chapter 7 takes a close look at changes occurring in the electric industry. Everything from fuels to business models is in flux. We will discuss the who, what, where, when, and why of change in electrical generation and consumption.

In Chapter 8 we delve into the transition to a more sustainable electric industry, offering suggestions for clean energy priorities and principles. Chapter 9 looks at the future of how we will power our planet, including advanced technologies that might emerge.

Finally, Chapter 10 is titled, somewhat fancifully, "Our Crystal Ball." We will not attempt to describe a comprehensive scenario of the future, but rather will list some highlights of what we think will happen in the building, transportation, and power production sectors, along with the roadblocks and obstacles that are likely to occur, such as the embedded energy problem and dependence on electricity.

## THE AUTHORS

Who are we? Why did we write THIS book? And why did WE write this book?

[Roger] I describe myself as a recovering politician. I was elected to the Austin City Council in the early 1980s and am proud to say that I left office undefeated and unindicted. I then worked as an executive manager for the City of Austin for 20 years, working on the topics of this book—buildings, transportation, and power. In 2010 I retired as General Manager of Austin Energy, the municipal utility for Austin.

[Michael] I am a professor at the University of Texas at Austin, where I hold the Josey Centennial Professorship of Energy Resources in Mechanical Engineering. A lifelong Austinite and the son of a UT professor, I have conducted research and taught on energy, environmental, and industrial topics for both UT and the private sector for more than two decades. In 2019, I took leave from UT to work as

Chief Science and Technology Officer for ENGIE, a global energy and infrastructure services firm headquartered in Paris, France.

We met when Roger was General Manager of Austin Energy and Michael was serving on the city's Electric Utility Commission. We immediately recognized that we shared similar visions for a clean energy future and a commitment to finding solutions to the climate change problems facing our world.

### *The Folly of Predictions*

We do not think any one of the scenarios will play out exactly as we describe. The best assumptions and computer models can turn out false, and misinformed guesses can be correct. Academics are not terribly good at predicting the future. Arthur C. Clarke said predictions about the future were often wrong because of either failure of nerve or failure of imagination.

We suspect the future will be some amalgamation of current and future technologies, and there will be different levels of achievement of sustainable and economic goals. While we expect many of our predictions to be wrong, we hope that pulling them together into one place and juxtaposing them against one another will be useful.

Arthur C. Clarke also famously said, "Any sufficiently advanced technology is indistinguishable from magic."[6] In that spirit, we will start this book with the basic principles shaping our future technological infrastructure. From there, we'll talk about how buildings, transportation, and the power grid are going to evolve into sentient-appearing machines. And we'll explore what it means to live, work, and move about inside robots. Think of it, if you like, as a magical journey.

# PART 1

# THE ENERGY EFFICIENCY MEGATREND

*The relationship between information, technology, and energy conversions will shape our future. The most important overarching technology megatrend of that relationship is the movement toward energy efficiency.*

Everything is energy. Energy is the gasoline we put in our cars and the electricity we generate to power our lights and computers. Einstein showed us that matter and energy are equivalent, so that means our clothes, chairs, food, and other artifacts of everyday life are energy too. Energy comes in a multitude of forms, and we can change them from one to another.

When energy shows up in current events and everyday conversation, it is usually a handful of fuels that get attention: coal, natural gas, petroleum, nuclear, and renewables (all bundled together). These are the five major fuels today, and we use them directly or indirectly to power appliances; move cars, trucks, and planes; heat and cool spaces; and perform any other purpose for which we wish to convert energy.

Unfortunately, the majority of energy we convert does not provide the energy services we desire. More than half of our energy conversions end up in the form of wasted materials we sometimes label as pollution, waste heat, or even wasted motion in getting to our goal. That abundance of waste is the starting point for improving the global energy system, and is one of our motivations for writing this book.

## ENERGY EFFICIENCY IS THE TECHNOLOGY MEGATREND

Efficiency, which reduces the energy required to achieve desired purposes, is a solution to the waste problem. Thankfully, it is the key technology trend for society. Our technological history is an ongoing optimization of the energy conversion from fuel into useful attributes, such as work and heat. From medieval waterwheels and levers to the most sophisticated modern robot, there has been a general progression of efficiency in energy conversions. *In fact, we can postulate that the purpose of technology is conversion efficiency: the efficient conversion of any form of energy from Form A to Form B.* This is the Energy Efficiency Megatrend.

This trend manifests itself in almost all applications, regardless of the purpose for which we are making the energy conversion. It is not limited to the "energy efficiency" technologies of more advanced light bulbs, sleeker cars, and building weatherization that have been a focus since the 1970s. Rather, it is a more general megatrend that cuts across all sectors of society.

In this chapter we will look at the historical articulation of this megatrend, the mechanism by which we are achieving conversion efficiency in technology, and consequences of this trend on the buildings, transportation, and power sectors of our economy. We will also examine the applicability of this trend to technological progress, particularly addressing the question of whether technological innovation is accelerating, and whether this acceleration applies to all technological development.

Technology continually improves conversion efficiency by increasing information intensity. The outcome of these trends is that the building, transportation, and power sectors will require less material, less motion, and less time; that buildings, transportation vehicles, and power grids will evolve into sentient-appearing machines; and that these sectors will increasingly converge through power and information flows. Our future technological progress will be a mixture of rates of change in the sectors due to the limitations of the overall trend. In particular, we will address the limited applicability of Kurzweil's Law of Accelerating Returns.

## Historical Articulation

Among the first to prominently describe the general energy efficiency trend in technology was Buckminster Fuller, in his 1938 book *Nine Chains to the Moon*. He coined the term "accelerating ephemeralization." To paraphrase Fuller, we are getting more productivity from less material in decreasing increments of time. He even said that the trend would mean "Eventually you can do everything with nothing."[1] While that idea is appealing, it will eventually bump up against thermodynamic limits.

Decades later, Fuller pointed out how tons of copper telecommunications cable in the Atlantic were replaced by a satellite that could sit on a desk. He enjoyed recounting how it took Ferdinand Magellan years to circumnavigate the globe in a wooden sailing ship and the way by which transportation progressively moved faster, until an astronaut orbited the earth in just over an hour. He told one of the authors (Roger) that the exponential increase in speed meant that we would soon develop some means of teleportation, since he considered it inconceivable that the exponential curve in accelerating speed would suddenly cease.[2]

Ray Kurzweil is another influential thinker on future technology trends. He introduced the "Law of Accelerating Returns" when he posited that technological change is exponential, rather than the slower linear change that is more familiar to everyday existence. He said that in the 21st century we should experience not 100 years of change, but the equivalent of 20,000 years of progress.[3]

Many are familiar with one manifestation of Kurzweil's phenomenon from the semiconductor industry, popularly known as Moore's Law. In 1965, Intel co-founder Gordon Moore observed that the number of transistors that could be placed on a silicon computer chip doubled during that year.[4] He noted, "certainly over the short-term this rate can be expected to continue." In 1975 he revised the doubling rate to two years, and later it settled into a time frame of about 18 months. This idea, borne out by reality, that computing power would double every couple of years is an evocative demonstration of accelerating technology improvements.

Kurzweil points out that Moore's Law was not the first, but rather the fifth paradigm in the exponential growth in computation. The computational curve he charted includes five different technologies: electromechanical, relay, vacuum tube, transistors, and integrated circuits.

Kurzweil and others have recognized that technology proceeds in what are called "S curves." That is, a new technology will often develop very slowly, then experience an exponential growth in

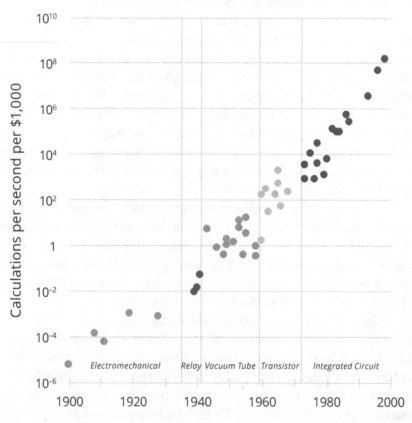

**A CENTURY OF EXPONENTIAL GROWTH IN COMPUTING CAPABILITY**

Source: Ray Kurzweil, *The Singularity Is Near: When Humans Transcend Biology*

**Figure 1.** Cost-normalized computing capabilities grow exponentially

efficiency and adoption, then level off and improve very slowly after an inflection point that ends the exponential rate of change. After that, another technology may develop that continues the efficiency trend for that particular function. This is what Kurzweil was describing as he noted all the different technologies that led to an overall exponential growth in computing power. As you would expect, he projects a new technology will continue that exponential growth after the limit of silicon chip technology is reached.[5]

Kurzweil also singled out solar technology as one that is experiencing exponential technological change along with an accompanying exponential drop in cost, and speculates that *all* energy technologies will progress at an exponential rate because "energy is information." He believes that the nanotechnology revolution will transform every aspect of energy.[6]

The impact of this rate of change is captured in the title of Robert Bryce's book, *Smaller Faster Lighter Denser Cheaper.*[7] Bryce shows how historically our most world-changing technologies have undergone this transformation: the printing press, vacuum tube, diesel and jet turbines, power plants, digital communications, and other important technologies have become "smaller faster lighter denser cheaper" at some point in their development.

The broadest application of the phenomena is suggested by the futurist James Canton in his book *Future Smart,* in which he makes the bold claim that "Every technology is doubling in processing power, storage and capability every twelve months."[8]

How is technology doing this? How are we consistently improving our different technologies to meet our workloads using less material, less energy conversion, less motion, and in less time? What exactly is happening across the different technologies to achieve ever-increasing conversion efficiency?

## INFORMATION IN TECHNOLOGY

We think the answer is that the *information intensity* of technology is increasing. We define information as physical order and note that, unlike Claude Shannon's formulation that established information theory, here the meaning of the information is important. Specifically, the information must be relevant to the function of the technology.

Let's review the basics: Everything is some form of energy, either in motion or as stored energy. Technology transforms energy, and that transformation is optimized by information. The more we know about the intended form of energy, the initial form, and the technology we use for conversion, the more efficiently we can accomplish the task. Thus, the information intensity of the technology and the application of the technology is key.

We have increased the information intensity of our technology in several ways: through the information content of the materials we use; through design and engineering of our tools; by adding sensors, actuators, and computational capacity to the technology; by advancing our understanding of how to apply our technology; and now by providing sufficient information for some technology to learn, adapt, and operate autonomously.

### The Information Content of Materials

Our materials have become progressively lighter, stronger, and more specific to the work function to which they are applied. Here, the atomic and molecular structure of the material is the information content. Whether it is the hardness of a diamond or the fluidity of water, it is this internal structure that makes the material useful. Historically, we have manipulated that interior structure through the science of metallurgy and innovation in the chemical and manufacturing processes, forming new materials to advance the construction and operation of everything from ships to buildings.

Such advances in structural information enabled the construction of skyscrapers as we advanced from stone to reinforced concrete and then to steel. Transportation vehicles progressed from wood and leather to steel and aluminum. Today we have developed advanced carbon composites for which the atomic structure is carefully defined to meet severe and specific stresses.

Material advances in the power industry have had a similar impact on the prime movers in electricity. Metal alloys allowed the manufacture of high-temperature and high-pressure steam turbines, increasing a power plant's thermal efficiency. Thus, a kilowatt-hour of electricity that required 7 pounds of coal to generate in 1902 only required 1.5 pounds of coal to generate in 1931.[9]

As we developed more sophisticated tools to first understand and then manipulate the molecular structure of our materials, we fundamentally changed the information content of our built infrastructure.

## Design and Engineering

The design of our tools has also become increasingly sophisticated. In the transportation sector, aerodynamics and fluid dynamics, along with lightweight, strong materials, lowered the resistance and increased the speed of our vehicles through air and water.

This trend of embedded intelligence also shows up in more precise engineering specifications, in everything from the simple hammer to sophisticated electronics. An original waterwheel could work even if the measurements were off by an inch or so from one wheel to the next, whereas modern turbines, aircraft, and similarly advanced technology must be engineered to fractions of millimeters and even nanometer specifications to work properly. For example, if the turbine's blade masses for a modern jet engine are out of balance or alignment, it won't work.

## Sensors, Actuators, and Computation Capacity

The addition of prolific sensors is more recent. Sensors give real-time information on the environment in which equipment is operating, increasing the precision of the energy conversion. This combination of sensors, actuators, and communication technology has led to the Internet of Things. Embedded sensors and microprocessors are making common objects such as light bulbs, doors, and microwave ovens aware and responsive.

We have actually seen rudimentary aspects of this type of on-board intelligence for many years. The efficiency of early windmills was greatly increased when Leonard Wheeler added a side vane and weight combination that would turn the windmill out of the wind when the force of the air became too strong and could damage the device.[10] By using more information—(e.g., which way and at which speed the wind is blowing)—the performance was improved. That approach foreshadowed modern wind turbines that feather their blades to avoid damage.

Modern wind-turbine blades morph their shape as they turn, forming an angle of attack with their airfoil to extract the most possible energy from the flowing air.

The original Wright brothers' plane was steered by the "feel" of the plane, whereas a modern commercial jetliner has thousands of sensors in the aircraft that monitor the "feel" of the aircraft. An advertisement for a General Electric jet engine boasts, "GEnx jet engines feature 23 sensors that measure 280 parameters continuously, generating 1 TB of data per aircraft per day. Data from the GEnx engine can be collected and analyzed in flight—detecting potential equipment issues and giving airlines the ability to manage issues before the aircraft lands."[11]

Finally, abundant and affordable computational capacity provides the means to translate the information gained from the sensors into focused action. Computers have given our technology decision-making capability, and now these capabilities are embedded in almost every appliance and much of our surrounding man-made environment. We no longer think of the microchip as an element of a computer, but as an integral part of commonplace "smart" appliances and devices, even if they are as simple as a toaster or teakettle.

## *Automation, Robotics and Autonomy*

Automation is taking responsibility for more of our everyday decisions, from cruise control and anti-lock braking systems in our cars, to automated doors and lights in buildings. The increasing information intensity of our technology has enabled a progression from mechanization to automation and then roboticization. When a technology has enough information from sensors and algorithms to perform a task repeatedly without human direction, it has achieved autonomy.

As artificial intelligence (AI) capabilities improve, large and complex systems may independently make decisions to meet their designated use based on their sensory perception and interaction with their human and natural environment.

Autonomy seems to be arising first in the transportation sector with driverless cars, unmanned aerial vehicles, and unmanned rail systems. We expect that buildings and the power sector will also develop into autonomous systems.

## Circling the Globe

Let's turn now to the classic example given by Fuller of accelerating technology: the increase in the speed by which man has circumnavigated the globe. It turns out to be a good example of how accelerating technology and the increasing information intensity (from new maps, notes on wind patterns, and so forth) facilitate a task—circling the globe—with less material, less motion, and in less time.

First, we need to understand that the basic task at hand—circling the globe—is governed and ultimately constrained by Newton's second law of motion, $F=ma$ (force equals mass times acceleration). We need to apply a force to a mass (ourselves in a vehicle) to achieve the acceleration necessary to circle the globe.

Now observe how technology has accomplished this through materials science, design and engineering, sensors, computers, and automation. Notice how the size and mass of the vessel shrank from the large sailing ship to an airplane to a small space capsule. And then note how the information embedded in the fuel changed, proceeding from the natural flow of the ocean currents and wind to a sophisticated, self-contained rocket fuel. Note also how the total motion was reduced by better information, and how the voyage changed from having no mechanical devices to complete automation. Finally, see how all this combined to dramatically reduce the time necessary to circumnavigate the earth.

The first people that we know to have circled the earth were the Spanish and Portuguese crew of Magellan who left Spain in five ships in 1519. One of their ships, the *Victoria,* managed to make it around the globe to reach Spain on September 6, 1522. It took them three years and a month.

The Victoria was a Carrack-type ocean sailing ship made of wood, about 66 feet in length, and weighing 85 tons. The 42-member crew navigated by the stars through mostly uncharted waters. There was no automation and limited information available beyond the landmarks the sailors could see with their own eyes. Wind and ocean currents provided the only propulsion. Meandering around the globe, they traveled about 42,000 miles. Only 18 crew members returned, the rest having deserted, died, or been killed—including Magellan himself.[12]

Although steam ships later improved on that time, the next big leap occurred on April 6, 1924, when the United States Army Air Service performed the first aerial circumnavigation of the earth. Four planes started the trip from Sand Point, Washington, and three completed it.

The Douglas World Cruiser was specifically developed for the attempt. The plane had a steel fuselage, strengthened bracing, and was 35 feet in length with a 50-foot wingspan. The 420-horsepower Liberty V12 engine was powered by burning refined petroleum. It weighed between two and three tons and had a two-member crew. The route was planned out with spare parts cached along the way.

Because more information was available from maps rather than simply the stars, and because they were not constrained by wind circulation patterns, the aircraft was able to travel directly in straight lines between airfields and followed a more northerly route, covering about 28,000 miles in total. The trip was completed in 175 days, and actual flying time was only 371 hours (vs. Magellan's 37 months), at an average speed of 70 miles per hour.[13]

Just 37 years later, on April 12, 1961, Russian cosmonaut Yuri Gagarin became the first human to circle the globe in outer space. The *Vostok 1* spacecraft, including its launcher, weighed just over five tons at launch and the space capsule itself weighed just over 2.5 tons. The capsule was a sphere 90.5 inches in diameter, composed of exotic metal alloys to withstand the temperatures of atmospheric reentry. There was not enough room for its one passenger to stand upright. The propellant was a mixture of nitrous oxide and amines. The spacecraft was completely automated, controlled by either on-board computers or from ground control. In fact, the pilot's manual controls were locked off, since there was uncertainty that Gagarin could operate them in weightlessness.

Multiple sensors measured key factors such as temperatures and humidity and the atmospheric pressure of the fuel tanks. Navigation was executed by a computer, which vastly eclipsed the information available to the ancient mariners of Magellan's day, but had a fraction of the computing power in a modern handheld phone. Once in orbit, *Vostok 1* took just 89 minutes to circle the earth.[14]

These examples of global circumnavigation illustrate Fuller's concept of accelerating ephemeralization. Since the fastest and longest travel before Magellan was by horse and a few ocean vessels, the increase in acceleration from the 1500s to the present followed a quickly ascending curve. This trend was achieved within the constraints of Newton's second law of motion by using information to dramatically reduce the mass of the vehicle, dramatically increase the force pushing it, place the vehicle in frictionless outer space, and take a shorter route.

## BUILDINGS, TRANSPORTATION AND POWER

Let's look at three consequences of the Energy Efficiency Megatrend and its particular impact on buildings, transportation, and the power grid. The first consequence is that these economic sectors will meet their functions with less material, less motion, and in less time.

### Less Material, Motion, and Time

We are using less material in construction and manufacturing. Dematerialization has become a common refrain, and new materials have become progressively stronger and lighter. These materials enhance the function of everything from skyscrapers to commercial jets. Buildings today soar many times higher than centuries-old structures, but with less material. Modern planes weigh less and go farther. We consequently move people and things faster and with less fuel.

Occurring in parallel with dematerialization, miniaturization is a long-standing trend in sectors such as electronics and telecommunications. Our technology has become smaller and now nanotechnology and quantum computing are pushing the physical limits of how small our tools and workplaces can be. The value of our goods has soared while their weight and size have diminished.

We are also achieving our objectives with less motion. For instance, energy conversion technology has moved from the linear, back-and-forth motions of an internal combustion engine to the more powerful rotary motion of a turbine, and from there to the more efficient power of the movement of electromagnetic fields. Electricity, or the movement of the electromagnetic field, is a prime example of achieving more power and productivity from less material and less motion. This is why the electrification of processes, from industrial manufacturing to transportation, improves energy efficiency.

Many processes can now be completed more quickly. Multistory buildings in China are erected in weeks. Amazon and their competitors are offering same-day delivery service. Many other services and products are moving as close to real time and on-demand as possible in response to a public's evolving preferences, including an expectation of quasi-immediate response and gratification.

### Sentient-Appearing Machines

We believe a second consequence of the Energy Efficiency Megatrend will be the evolution of buildings, transportation systems, and the power grid into sentient-appearing machines. We are not saying that we believe machines will become conscious. They may or may not; we really have no way of knowing. What we are saying is that we will increasingly interact with things in our environment—buildings, vehicles, and other things—*as if* they were sentient.

We are developing our technology with sentient-appearing interfaces. We talk to Siri on our iPhones, get in our cars and take directions from our GPS, and ask Alexa which team won the big football game last night. The use of natural language and gestures is an efficient way to interact with our technologies compared with manually engaging with our devices to get the information we want.

The science fiction future is full of examples. Sarah, on the television series *Eureka,* manages the house—and, in fact, *is* the house. HAL, the computer in the movie *2001: A Space Odyssey,* makes its own interpretation of how to fulfill the space mission. In another TV series, *Knight Rider,* the car has a mind of its own. As time moves on, we grow more comfortable just talking to our environment and expecting it to respond. We see this trend as being greatly magnified in the future as we communicate with the Internet of Things in a casual manner. Indeed, in the future if we speak to something and it doesn't respond, we will probably think it is broken.

Sentient-appearing buildings will manage the vast preponderance of our normal energy requirements at home and work, both managing consumption and generating and marketing electrical power, as well as becoming our assistant in many daily chores. A sentient-appearing transportation system will always know where we are and when we need transportation to another location. It will always be "on," and if we allow it, will listen to our conversations and determine what transportation services we need as we go about our daily lives. Power production systems for generating electricity or providing liquid fuels will autonomously keep the lights on and the tanks full. As creepy as it may sound, in the future we will be living, working, and moving about inside robots. This reality will be greatly enabled by and consequential for modern energy systems.

Although it is tempting to envision a single autonomous system operated by a single artificial intelligence scheme, it is more likely that intelligence will be distributed among nodes of the system, with different levels of AI appropriate to the function of the technology. It is more efficient for an "end use" system to gather, analyze, and act on certain types of data rather than have all the elements of buildings, transportation, and power responding to a centralized command and control system. Some end users may have minimal intelligence to receive instructions and perform their functions, whereas other artificial intelligences may control and coordinate large parts of our technology infrastructure. Some observers refer to this phenomenon as edge computing.

## Convergence

A third consequence of the Energy Efficiency Megatrend is the increasing exchange of power and information among buildings, the transportation system, and the power system. We currently have an energy system where electrical power is produced by large, mostly fossil-fueled power plants, then pushed over a one-directional transmission and distribution system to buildings, and a transportation system that is running almost entirely on petroleum and almost completely disconnected from the electrical system.

We are transitioning to a new type of system where utilities are producing power from a larger diversity of fuels. Buildings are becoming more energy efficient and starting to produce power on-site. The transportation system is finally transitioning from petroleum

# OUR ENERGY FUTURE

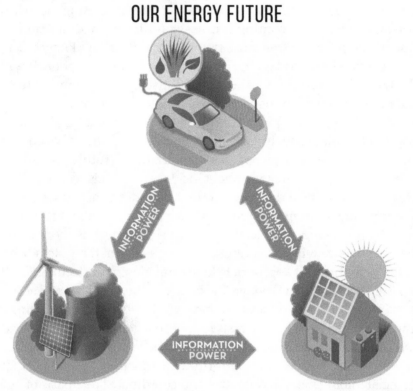

**Figure 2.**

to a variety of non-petroleum fuels. The one-directional flow from power plants to buildings is becoming bidirectional flows of power and information connecting all three sectors: building, transportation, and power production.

This exchange of information and power between the sectors is growing, in some cases exponentially, and changing the fundamental relationship between the sectors from what we have historically experienced. In many important ways, the functions of the three sectors are converging into a single intelligent technological system.

Buildings are becoming power plants, and in some instances are becoming mobile: mobile homes, offices, clinics, libraries, etc. The transportation sector is now plugging into buildings and utilities for electric fueling and acts as a capacitor for the electric grid.

This system is increasing in size, power, complexity, and intelligence. The lines of distinction between the three sectors are blurring, and the efficient evolution of each sector is becoming increasingly dependent on the others.

It will be interesting to see if future experience shows that this system is starting to display characteristics of a *dissipative structure*, whereby a form is being maintained by the intake and dissipation of energy. In other words, are buildings, transportation systems, and the power grid starting to act as a single integrated, thermodynamic system?

Fuels and power flows can be converted throughout the system, ranging from large centralized power plants and wind and solar "farms" to distributed energy resources, such as solar panels, and down to the micro energy harvesting of minute power flows of movement, light, heat, and chemical reactions. Sun, wind, and vibrations impacting building sites, vehicle fuels, and motions—all of these energy flows can be converted and integrated into a very complex energy system.

Most of the power will be converted, consumed, stored, and dispatched from building sites rather than conventional power plant sites. Some buildings with large surface areas for solar and infrequent use, such as stadiums, parking garages, and storage units, will primarily be power plants. Most urban buildings will be multistoried residential and commercial facilities, and although they will probably not be net zero-energy, a significant portion of their energy needs will be met on-site.

The overall system can operate at very high efficiency, with technological advances ensuring that energy is converted only when needed, stored efficiently when not needed, and consumed as efficiently as possible. The system can also match fuels to workloads, whether electrical or thermal, significantly reducing loss from transmission and distribution and reducing the number of energy conversions.

This convergence is taking place because it is more efficient for each sector to meet its function by working and integrating with the other two sectors. For example, the transportation sector becomes more efficient by communicating with the buildings in its environment and integrating with an intelligent fueling system, such as the electric grid. Each sector has similar advantages from technological collaboration with the other parts of the system.

## RATES OF TECHNOLOGICAL CHANGE

The final question we want to address is whether the pace of technological advancement is accelerating as predicted. Are we headed toward a technological "singularity" where the pace of change will exceed our ability to comprehend it?

This is an important question. If technology in general is advancing in the fashion some futurists have claimed, then frankly all our resource problems—climate change, food, fresh water, poverty—will be solved, and soon. To be clear, that sounds improbable, and other moral or ethical problems would remain, but making progress on our resource challenges is an exciting proposition nevertheless. So let's take a closer look at this "technological optimism" that claims technology is doubling in capability every few years, and will soon be advancing fast enough to address our global economic and environmental concerns before they become catastrophic.

### *Limits on Accelerating Technology*

While we are optimists ourselves, there are a few reasons why unquestioning faith in our innovative abilities may not be warranted. First, there are constants of the physical world that will strictly limit what future technological innovation can achieve, including physical laws that govern nature's processes. For example, although much of our technology is, indeed, becoming smaller and lighter, nanotechnology is the bottom limit, as we eventually collide with Heisenberg's uncertainty principle when we try to shrink anything further. And the speed with which we can send a message from one part of the globe to another is really not going to change, because we are already communicating at nearly the speed of light, which serves as nature's speed limit.

We have no indication that accelerating technology can change the laws of physics. Indeed, it is our deep understanding of these laws that underlies advancement. Artificial intelligence may show us nuances that we have not imagined, and even greatly expand what we thought possible within nature's constraints, but we would not expect AI to start producing perpetual motion machines or faster-than-light spaceship drives.

Resources can also be a limiting factor. Although there is much innovation in combining elements in novel ways, no new atoms are being created on earth and the useful combinations are not endless, so minerals available in the Earth's crust underpin global supply chains. The modern smartphone contains almost half of all the elements known on the planet, and some batteries and other power electronics depend on rare-earth materials that are in limited supply. At the same time, rare-earth materials are not as rare as their moniker implies. However, their mining capacity might be limited or in regions of the world that do not protect human rights nor abide by western democratic interests. Other advanced batteries and technologies under development avoid these problems by using different materials.

Another reservation is the applicability of exponential acceleration in technology. Although we believe that the Energy Efficiency Megatrend applies to all technology development, that does not mean that every technology is advancing at an exponential rate of change. We do not think that Kurzweil's Law of Accelerating Returns, or Fuller's "accelerating ephemeralization," are applicable to all technologies at all times.

Currently these laws seem to be applying to information and communication technologies. We should not underestimate the impact this can have, as it can make almost any other technology sector "smarter" and more effective in meeting functions. But this is not the same as all technology experiencing exponential growth.

For a multitude of technologies—buildings, ships, planes, trains, automobiles, power plants—none are doubling in capacity or reducing costs by half every twelve months, even though incremental improvements in capacity and costs continue, and almost all of them are being profoundly transformed by the digital revolution.

Kevin Kelly examines the accelerating technology trend and exponential growth in *What Technology Wants* and makes an important point regarding applicability.[15] He notes that the examples of this trend are all about scaling down and working with the small. He notes that our new economy is built around technologies that scale down well and use little energy per device.[16]

He points out that you do not find the same exponential effect if scaling up to bigger technologies, like buildings and planes. Airplanes are not improving in capacity, speed, or efficiency at exponential rates.[17] Rather, their progress is slow, methodical, and incremental. In the timespan that chip designers will double transistor density and halve the cost, plane designers will gain less than 1 percent improvement over the latest models.

This reality of the challenges of scaling up is relevant to all three sectors discussed in this book—buildings, transportation, and power. When these three scale up, the energy requirement scales up just as fast and is a major limiting factor.

As we said in our article, *Four Technologies and a Conundrum: The Glacial Pace of Energy Innovation,* there has not been a fundamental innovation in the conversion of primary energy into motion or electricity in well over a century.[18] The entire global economy essentially runs on four primary energy conversion technologies—the steam turbine, combustion turbine, spark ignition engine, and compression-ignition engine—invented in the late 1800s or earlier. This realization is important because computers, digitization and Moore's Law, all of which are about rapid evolution, are really just increasingly sophisticated ways of manipulating electricity. But the underlying electrical generation technologies are evolving slowly.

The steam turbine was invented in 1884; the combustion turbine in 1791; the first hydroelectric turbine operated in 1878; the wind turbine in 1888; and the first solar photovoltaic (PV) cell in 1883. Even space-age fuel cells were invented in 1839. While the use of nuclear fuels is relatively new, with the first power plant demonstrated in 1951, it is just one more way to boil water for a steam turbine, which is over 130 years old.[19] The rest of our energy consumption is for standard devices like boilers, burners, water heaters, and cooktops.

Although these technologies have improved in efficiency (at an incremental rate far more slowly than suggested by Moore's Law), they were essentially mature technologies within a couple of decades from their invention. There has not been a fundamentally new way to generate electricity in more than a century that has captured significant market share.

Vaclav Smil calls Moore's Law "Moore's Curse" because it creates false expectations for technological progress, and he has done extensive analysis on the rate of change in fundamental technologies—especially energy technologies—that show Moore's Law is not applicable.[20] He noted, "More importantly, for some basic energy production processes and conversions…there have been either no, or only marginal, gains in the best performance or in maximum ratings and unit capacities during the past four decades."[21]

Because change is taking place in renewable energy sources and not fossil fuels, Smil looked specifically at wind generators and photovoltaic panels that convert solar radiation into electricity. He noted that while wind turbine towers and blades have increased in size dramatically, increasing output and reliability, their efficiencies have remained largely unchanged since the late 1980s, at around 35 percent, and it will be impossible to double their efficiency again due to thermodynamic limitations.[22]

The first solar cell, which was invented in 1883 by Charles Fritts, could convert about 1 percent of the sunlight striking the cell into electricity.[23] Today, after more than a century-plus of effort, the efficiency of commercially available PV cells generally remains below 25 percent. In labs, the efficiency of PV cells had risen to only 47.1 percent in 2020.[24]

And it is not just certain technological sectors that stall. Even though some technology seems to be changing rapidly, and the level of innovation and patents are increasing, one has to question the impact and importance of the innovative advances. An increase in the number of research publications and patents does not necessarily mean a dramatic change in our economy or quality of life.

In *Creating the Twentieth Century: Technical Innovations of 1867–1914 and their Lasting Impact,* Smil concluded that the fundamental technological innovations shaping our modern world were generally invented between 1867 and 1914, in what he calls the Age of Synergy.[25] During this period, we saw the invention of the generation, transmission, and utilization of electricity; the internal combustion engine; the automobile and the assembly line; the airplane; telephone, telegraph and wireless communication; the Haber-Bosch process to produce ammonia for fertilizer; prestressed concrete and steel skeleton skyscrapers; crude oil tankers; X-rays; movies and radio broadcasts; vacuum tubes, reliable lighting, air conditioning, and much more. Smil believes that the most inventive decade in history was the 1880s.[26] The 1960s, when most of the technology underlying information technology and the internet were invented, only to take root decades later, is the closest parallel we have in recent history.

The final reason that all technology is not advancing exponentially has to do with the external factors affecting technological development. So far we have only been talking about technology innovation. The innovation and adoption of new technology does not occur in a vacuum. Economic conditions, environmental changes, legal structures, political entities, research and development budgets, and cultural norms affect technological innovation and adoption.

Alvin Toffler pointed out how different aspects of our society move at different speeds, which he called "desynchronization." Technological change moves fastest, while political and policy change move slower, and the legal system slowest of all. This mismatch in the rate of progress in different aspects of our society and economy can inhibit progress.[27]

Decision-making in the government sector, for example, or the pace of the legal system, seems to be slowing. Despite our many technological advances in the last seventy-plus years, infrastructure projects like pipelines require much more time. It takes a decade to get a permit today for a major pipeline; in the 1940s, the BIG INCH transcontinental pipeline was permitted in days and built in months as part of the military mobilization for World War II.

So it is worth distinguishing between areas of technology that will be subject to periods of exponential growth in capacity and the major areas of our modern world that will change slowly, if at all. We need to be mindful in our analysis of where and how the exponential growth within certain technologies like genetics, nanotechnology, and robotics will transform our landscape, and where certain fundamental power needs will still be required and will not be transformed. The intertwining of these separate trends will make prediction of future energy scenarios difficult.

These different rates of change may mean that progress in some sectors won't be what we expected, resulting in a technological future that may seem both advanced and not changed at all. Here are two examples:

### Transportation Future

Consider transportation vehicles. The exponential rate of change in information and communication technology means that we can expect future vehicles to be "smart," autonomous, connected, and even sentient-appearing. The transportation sector as a whole may be interconnected to the point where it is always aware of where we are and where we intend to be, according to our schedules. In fact, with apps on our smartphones tracking our locations and reservations, and with our phones speaking to our cars via Bluetooth, that might already be the case. It will also know the schedule and navigation for travelers around us. If we wish, we might be able to just declare out loud where we want to be at some point in the future and expect the transportation system to immediately present us with options for travel and arrange whatever combination of vehicles necessary to deliver us to our destination at the appointed time. Along the way, it would nimbly avoid accidents by communicating with other vehicles. Stoplights might become archaic and obsolete, as cars would simply weave through each other at intersections, as if by magic.

We also expect the continued electrification of our transportation system due to increases in the power and storage capacity of batteries. However, since their first demonstration in 1800, batteries have improved in a linear progression, not exponentially. That means that over time batteries are improving enough to electrify our cars and trucks, and even support air taxis and short-haul aviation. But linear progression in battery technology is unlikely to provide the order or magnitude change in power and reduction in weight necessary to propel large commercial jetliners or ships across the country and over the oceans.

And the exponential increase in speed that Fuller used as an example has stopped. The time needed to circle the globe, or even travel from New York to Los Angeles, has not increased exponentially, or even at a slower linear rate. It takes about the same amount of time to travel from New York to Los Angeles as it did in 1958. As Smil noted, "The speed of intercontinental travel rose from about 35 kilometers per hour for large ocean liners in 1900 to 885 km/h for the Boeing 707 in 1958, an average rise of 5.6 percent a year. But that speed has remained essentially constant ever since—the Boeing 787 cruises just a few percent faster than the 707."[28]

There will be reductions in that time as we move to supersonic aircraft and even suborbital rocket travel, and shorter trips may be faster due to magnetically levitated trains in vacuum tubes, but that will still not exceed the speed that Yuri Gagarin achieved in 1961. And even those incremental advances are decades away. Clearly, the exponential increase in transportation speed has stalled.

Note that it is not the case that a particular technology has completed an S-curve, and that the *function* is then handed off to another technology to continue the exponential growth. In this example, the function—increasing the speed of the physical movement of people and goods—has stalled for decades.

Let's now look at a second example: the technological progress of solar power.

### Solar, Moore's Law, and the Missing Exponential

It has become a popular idea that solar power will follow the exponential curve that the computer industry experienced due to Moore's Law, exponentially increasing capability as costs plummet. According to this viewpoint, future energy will be extremely cheap due to solar power; all buildings will be net zero-energy; and "poles and wires" will go away (this last claim usually made while waving a cell phone).

Although some important parts of that vision ring true, other parts do not. It is reasonable to expect that solar power will become extremely cheap and ubiquitous, with solar panels covering many buildings and surfaces. However, the comparison of solar power to the computer industry is flawed in one important aspect: there is a missing exponential rate of change.

When we look in more detail at the remarkable growth in the computer industry, we see three different exponential rates of change that have taken place.

First, there is Moore's Law. Since the number of transistors on the chip corresponded to the calculation capacity of the chip, Moore's Law meant that the *computing power* of the computer chip doubled roughly every 18 months. This prediction's prescience is legendary, as the path of increasing computational capability has, indeed, followed Moore's Law for roughly 50 years.

This astonishing growth in computer power has fostered the rise of personal computers, tablets, smartphones, and microprocessors embedded in our everyday appliances that dwarf the early computers in computational power.[29] The computers in your family car are cheaper and smaller and much more powerful than the guidance system that put Americans on the moon.[30]

Moore's Law is important, but there are two other important exponential rates of change that have enabled the phenomenal growth of the computer industry. The second key trend is the exponential growth in the manufacture of computer chips and components. What started as a very expensive small batch of original chips suddenly leapfrogged to millions and then billions of chips through automated manufacturing processes. Not only had the computer chip's computing power grown enormously, but also we suddenly had billions of them.

The third exponential rate of change was the decline in the price of the computer chip. With the capability to manufacture billions of these components using relatively cheap ingredients, the price of computers declined in real terms at the same time that their capabilities increased.

These exponential rates of change combined to provide us with the phenomenon that we have become familiar with for many years: the new computer on the market is twice as fast, with twice the information storage, as the one I just bought a few years ago, but at a lower price.

Now compare this story to the evolving trajectory of the solar cell industry. It has the second and third exponentials, but not the first. There have been exponential improvements in manufacturing processes, leading to an exponential increase in the total installed capacity of solar. And due to advances from materials science and designing and engineering, the cost of manufacturing a solar cell has dropped exponentially over the last thirty years, reminiscent of the falling costs of microprocessors.

These two exponential rates of change mean that solar power is going to be common and cheap. That means solar power can meet a substantial portion, even most, of our energy needs.

However, the first exponential rate of change in the power of the device—the real Moore's Law—does not apply to solar. Whereas the computing power of a computer chip has been doubling every 18–24 months, the electrical power that comes from a solar cell has not been doubling exponentially, and cannot.

This constraint exists because a solar cell is just an energy conversion device for solar radiation. It can never generate more electricity than the amount of energy contained in the sunlight unless supplemented by an external fuel. That upper limit is determined by nature, not engineering. A square meter of solar cell will never generate more than 1,366 watts of electricity (the maximum amount of energy conversion from solar radiation), and, in fact, solar cells are not getting anywhere near their technical potential.

If solar cells had been following Moore's Law in terms of exponential growth in electrical power, then a square meter of solar cells today would be producing over a million watts of electrical power. Although there have been jumps in conversion efficiency, the constraints of the Laws of Thermodynamics mean they will never convert more than 100 percent of the sunlight. In reality, today's solar cells generate less than a fourth of that. So instead of a million watts per solar cell, we have only a few hundred.

Eventually, photovoltaic (PV) technology may be replaced with carbon nanotube antenna technology, which can theoretically achieve 80 to 90 percent conversion efficiency.[31] That would, indeed, be another doubling in power output and halving of costs. However, that would be it. We run into the physical limit of how much energy can be converted from sunlight, and the power output will be capped.

Although solar will be covering a lot of our buildings and surfaces, and will, indeed, meet most of our power needs through a combination of utility scale solar farms and distributed photovoltaic generation, the missing exponential on solar power explains why two recurrent visions—"All buildings will be zero-energy or positive energy buildings" and "Poles and wires are going away"—are actually myths.

In most climates, the available surface area on single-story residences and commercial buildings is sufficient to enable zero-energy operation because they can be that efficient. However, multistory buildings with any significant electrical load, whether residential multifamily, commercial, or light industrial, will generally be unable to meet their energy requirements by converting the solar energy that falls on the site. There is simply not enough primary energy to convert to electricity, no matter how much better solar panel manufacturing becomes.

There certainly have been some successful multistory commercial zero-energy buildings (ZEBs). Most notably they include the National Renewable Energy Laboratory, the new buildings for Rocky Mountain Institute, and the new Living Challenge building for the Bullitt Center in Seattle. However, even these buildings are limited in the energy loads they can meet through on-site generation and storage. They are not going to host a data center or meet industrial electrical loads, and they are vulnerable to changes in solar access. If tall skyscrapers were to be built on adjoining land, these buildings could quickly lose their ZEB status. They could remain 100 percent renewable-powered buildings, but the renewable power would be transmitted through the electrical grid and not generated on-site.

There are some multistoried structures that may, indeed, become positive energy buildings (generating more energy than they consume) because they have a large surface area but small or nonexistent electrical loads. These include parking garages, reservoirs, warehouses, and other buildings that may experience seasonal loads such as schools and stadiums. Many of these buildings could be covered with solar panels and meet all or most of their energy needs most of the time, or at least seasonally, and they would also essentially become power plants for the electric grid most of the time.

However, most residents of our large cities will not be living or working in a zero-energy building. Building codes that mandate zero-energy will not be achievable in most multistory commercial or multifamily residential structures. Perhaps a better approach would be to maximize efficiency in the structures, calculate the potential for on-site generation, and then maximize that generation, with the remaining loads being met by off-site renewable energy delivered through the grid.

The prevalent second myth is that "poles and wires are going away." People have the mistaken notion that the power supply is going the way of the telephone "land line." Indeed, wireless communication has replaced telephone lines to a large extent.

But beaming kilowatts of electricity through the air from building to building remains a vexing and unsolved challenge. To date, a physical conductor is required. (Smaller amounts of electricity can be moved without wires through induction, allowing some appliances and even cars to eventually be charged in this fashion at a short distance.)

So even if we cover all the buildings with solar panels, we must physically connect those buildings if we want to move substantial amounts of electrical power between them. In fact, with a massive deployment of distributed generation and storage, there may even be more poles and wires than we see today.

It should be noted that in *The Singularity is Near,* Kurzweil used solar panels as an example of exponential growth and proof that our power technology is, indeed, growing exponentially. However, he referred to the *cost* of solar dropping exponentially, and not the increase in power output. While certainly good news, those plunging costs are the consequence of other underlying phenomena such as increased global silicon mining capacity, improved manufacturing techniques, and economies of scale.

In summation, the Energy Efficiency Megatrend governs our technological progress, and information is the crosscutting attribute that facilitates it. We will see our technology getting smarter and attaining autonomy in most of our economy. We will be living and working inside sentient-appearing buildings and moving about inside a sentient-appearing transportation system. It will all be powered by a unified energy system that will autonomously keep the lights on and the vehicles fueled. But that does not mean that the future will be changing so fast that we cannot comprehend it, nor will all technology be changing rapidly. We may be living in a magical land of enchanted objects in the future, yet still using some technologies that our grandfathers would easily understand.

Let's proceed now to look at the future of each of these sectors. We will see how we are currently in the "smart" age: smart buildings, smart cars, smart grid, etc. And we will see how that increase in embedded intelligence will help make our economy more sustainable. Finally, we will watch the evolution of our buildings, transportation vehicles, and power grid into sentient-appearing machines.

## CHAPTER 1 SUMMARY

1.  The Energy Efficiency Megatrend will govern technology trends in the future.

2.  Greater conversion efficiency will be accomplished by increasing the information intensity of technology.

3.  One consequence of this megatrend is that buildings, the transportation sector, and the power grid will meet their functions with less material, less energy conversion, less motion, and in less time.

4.  A second consequence will be the emergence of buildings, transportation vehicles, and the power grid as sentient-appearing machines.

5.  A third consequence will be the convergence of buildings, the transportation sector, and the power grid into a single system with integrated power and information flows.

6.  Finally, there will be different rates of technological change occurring in these sectors, with intelligence and awareness rapidly advancing while many other aspects of the technology change at much slower rates.

# PART 2

# THE FUTURE OF BUILDINGS

*The bedside alarm sounded its usual aggressive tone and Bob stumbled out of bed and made his way to the bathroom, hungover from the night before. He stared at his reflection sullenly before turning on the faucet and washing his face.*

*As the coffee brewed, Bob turned on the television in the kitchen and tuned in the local morning news broadcast. The newspaper, on the front porch, could wait.*

*Toast, scrambled eggs, bacon, the usual. Bob moved around the kitchen by rote, pausing occasionally to check his email. He recorded a couple of phone memos having to do with his sister's wedding. He was in charge of planning the rehearsal dinner and there seemed like a million details to manage.*

*After breakfast, he brushed his teeth, dressed, and ordered a car to take him to work. With any luck, by the time his driver arrived, the black coffee and aspirin would begin to make him feel a little more human.*

# BUILDING TRENDS

For much of human history, wood was not only a primary source of construction material, but also one of the few fuels available for heating, cooking, and lighting. Energy use in the home was largely devoted to those three endeavors.

But a revolutionary development in Menlo Park, New Jersey, in December of 1879, signaled an epochal shift in energy use. On that date, Thomas Edison lit up his home and workshop with hundreds of electric light bulbs and ushered the world into the new age of electricity. For the first time in human history, there was a source of artificial light that did not depend on burning wood, whale oil, candle wax, natural gas, or kerosene inside the home.

Electric appliances soon followed, as did air conditioning, which quickly evolved from a desirable luxury to a virtual necessity in just a few decades. Widespread adoption of air conditioning accelerated population shifts to hotter and more humid regions, such as the coastal U.S. and the desert southwest, as well as other hot and humid regions around the world.

If the story for the last century was rising energy consumption for our buildings, the driving factors for the future will focus on gains in energy efficiency. Buildings over the next few decades will not only become more energy efficient, but will also be better constructed, smarter, and more automated.

In this chapter we will see how the Energy Efficiency Megatrend will lead to more efficient and smarter buildings. There will be robots at every level, starting with design and engineering, continuing through construction, and ending in the operation and maintenance of the building.

Sentient-appearing buildings will not be part of the near future, but buildings will become smarter through a vast array of sensors and intelligent appliances. We will also see the convergence of buildings with the power grid through connections to the "smart grid," the proliferation of solar and other forms of generating power, and energy storage located within the building. Finally, the transportation sector will become more integrated with the built infrastructure through charging stations and vehicle-to-home technology. In the near future we will see the age of the "smart" building, a giant step toward the sentient-appearing, autonomous buildings of the far future.

## BUILDINGS WILL BECOME MORE ENERGY EFFICIENT

Over the next several decades, expected growth in the energy loads of buildings will be small due to energy efficiency gains. Scenarios from both the U.S. Department of Energy (DOE) and the International Energy Agency (IEA) depend heavily on technology to reduce future consumption.

### *Residential Energy Consumption*

Although the average energy and electricity use per household in the U.S. is declining, we are building more homes, and the houses are getting bigger. The average floorspace per house is projected to increase from 1,786 square feet per house in 2019 to 1,987 square feet in 2050[2]. So, despite the continued trend of improved energy efficiency per household, the EIA projects that purchased electricity by the residential sector will slowly grow by 0.6 percent per year.[3]

Increased demand for space cooling and electronic equipment such as security systems and rechargeable devices is expected to slightly offset energy intensity reductions in lighting, space heating, and some appliances such as televisions and personal computers.[4] Due to the lighting efficiency requirements of the Energy Independence and Security Act of 2007, which prescribed performance targets for light bulbs, lighting energy requirements in 2050 are projected to be 40 percent lower than in 2019.[5]

Space heating loads are also expected to continue dropping due to more efficient heating, ventilation, and air conditioning (HVAC) systems as well as population shifts to warmer climates.[6] However, the movement of people to hotter climates means air conditioning loads should increase, taking the place of the missing heating loads. EIA projects that air conditioning will grow more than any other residential building end use, and that growing demand for air conditioning, electronics, and water heating, will more than offset declines in heating and lighting energy consumption.[7]

## Commercial Energy Consumption

The EIA projects that commercial floorspace will grow 1 percent per year through 2050, with the largest growth occurring in the energy intensive sectors of health care and lodging. However, the electricity consumption per square foot declines at an average of 0.2 percent per year, with declining electricity intensity in lighting, refrigeration, space cooling and heating, ventilation and water heating.[8]

Lighting accounts for the largest decrease, falling by more than 2 percent per year. Efficient LEDs are expected to displace linear fluorescent lighting as the dominant commercial lighting technology by 2030.[9] However, these declines are more than offset by other electrical uses, such as office equipment (not including computers), whose energy intensity increases by 1.6 percent per year.[10]

A significant part of the new demand for electrical growth is occurring with data centers. Their electric load has been doubling every four years, causing concerned analysts to worry it could triple within a decade. However, energy efficiency trends within data centers have caused energy consumption to grow much more slower than the demand for computing power.[11]

## SMARTER, FASTER CONSTRUCTION

Starting with the goals, preferences, and realities of a project and its participants, the company Autodesk Research, and their Project Discover, uses a computer to morph and meld parts of thousands of building designs, considering a multitude of human and physical variables, while learning and changing with each project.

Autodesk Research refers to this process as generative design. This human-fed artificial intelligence offers the time and resource efficiency of old school pre-fab, while creating building designs completely unique to each project.[12]

The engineering of structures will also become much more precise. Such precision, enabled by computer design and robotic assembly, will speed up building construction and improve quality and dependability, while simultaneously requiring less work. In Japan, steel I-beams are cut by robots to a tolerance of 0.16mm, which is incredibly precise by traditional standards.[13]

As a consequence of these innovations, the construction time required for erecting buildings has become relatively short compared with historical timelines. Habitat for Humanity constructed a house from pre-manufactured parts in 3 hours, 26 minutes, and 34 seconds in Shelby County, Georgia. That beat their previous record of 3 hours, 44 minutes, and 59 seconds set in 1999 in New Zealand.[14]

Broad Sustainable Building, a Chinese construction company, built a 30-story building in just 15 days,[15] and a 15-story hotel in just six days.[16]

Synergy Thrislington built a basic 10-story building structure in India in 48 hours in 2012.[17] Inspired by Chinese construction times, Forest City Ratner, a development company in New York, is prefabricating apartment modules that can be lifted and fitted into a modular high-rise construction project in as few as 12 minutes.[18] Overall, because of saved time and avoided energy use, this rise of robotics is closely coupled with energy efficiency.

### Construction Robots

Accelerating things further, robots will become prevalent throughout the construction industry for building and operating our structures. Though a rarity at present, it shouldn't be long before we see robotic operation of most traditional construction equipment. Advantages include faster production times, finer precision work, and greater safety for human workers, but they also mean a replacement of muscle power with electrical power.

Robots have been developed for demolition work, conducting inspections, sorting construction waste, painting walls, finishing concrete floors, welding frames, and conducting layout and measurement tasks. There also exist trucks and heavy machinery on construction sites that can be remotely controlled.[19]

The University of Sydney developed an automatic excavator that could eventually lead to the development of larger, robotic-operated construction equipment. Smaller construction robots are already being used for specific building tasks. In Japan, the Kajima Company's "Mighty Hand" robot lifts heavy concrete walls and other large materials. Takenaka's Surf Robo automatically compacts concrete floors, and the WR mobile robot performs column-to-column welding operations.[20]

Most robotic assistance occurs today in factories creating pre-fabricated building parts, such as insulation and windows, and their duties should expand over time. Specialized robotic systems in development will construct walls and take over other aspects of building construction. Baltimore's Blueprint Robotics combines workers, robots, and machines to piece together walls and roof that arrive at the worksite already 60 percent assembled.[21]

At the same time that advances in hardware are allowing robots to handle more on-site construction tasks, new software will automate entire construction processes. In the future, we will simply connect architects' schematics to materials handlers and construction robots on-site, rather like ordering a pizza online with options for different toppings or other ingredients and special requirements.

## Robots and Employment

One important society-wide impact to keep in mind is the replacement of many construction workers by robots. Similar outcomes will occur as the transportation and power sectors evolve.

In the past, the economy changed with automation and jobs moved to different sectors. Workers moved from farms to the city as agriculture mechanized, and then to the service sector as food production and manufacturing became increasingly automated. Now we see automation moving into almost every part of the economy.

The latest response to this issue suggests that workers and robots will work together, especially as artificial intelligence (AI) takes a foothold in society. The book *Humans + Machine, Reimagining Work in the Age of AI* by Paul R. Daugherty and H. James Wilson relates numerous examples of what the authors call the "missing middle." In this vision, humans complement machines and artificial intelligence gives humans superpowers. Together, humans and machines increase productivity.

Robots reduce the risks that have traditionally faced human workers in the construction industry, and more factories and workplaces are finding ways for the worker and the robot to split appropriate tasks to increase productivity.

We certainly think a lot of that is already occurring. Whether it will offset job losses due to automation is yet to be seen. The workers who are displaced are not usually the ones who get new jobs arising from automation. And if the pace of automation is accelerating, then job retraining becomes very difficult in an aging population.

We are not experts in this field, but in the economic sectors addressed in this book—buildings, transportation and power—numerous jobs, from construction workers to truck drivers, are going to be displaced by automation. It's in the best interests of society to start addressing this issue immediately.

### 3D Printers

3D printing, sometimes called additive manufacturing, is going to completely transform the building construction industry. This technology will allow fairly rapid manufacture of modular construction elements for quick assembly, to exact specifications.

At the 2018 South by Southwest Interactive Festival in Austin, Texas, the charity organization New Story, along with robotics construction company ICON, unveiled the first fully permitted, up-to-code 3D printed house. The concrete-based home measured 800 square feet, cost $10,000 to build, and was completed in 24 hours.[22] Eventually, ICON plans to bring construction costs down to $4,000 and use their Vulcan printer technology to help ameliorate the global housing crisis.[23]

Using a massive 3D printer, a Chinese company "printed" its own 10,000-square-meter facility that itself is now being used to rapidly print highly affordable homes made of recycled material.

In a process they call contour crafting, WinSun Decorative Design Engineering Company completed 10 homes in a single day using a 3D printer 490 feet long, 33 feet wide, and 20 feet deep. These 200-square-foot homes are estimated to cost less than $5,000 each. What's more, each of WinSun's new homes was created using recycled construction materials and industrial waste and tailings. WinSun is planning to build more than a hundred new factories to "collect and transform" recyclable construction materials—chiefly construction waste—into materials for new, printable homes.

As if that were not enough, WinSun reports that it has also printed a 5-story residential apartment complex and a 1,100-square-meter villa. Each villa costs approximately $160,000 to print.

Homes like WinSun's constructions can be predesigned to accommodate ductwork, appliances, wiring, plumbing, air conditioning, and more, so that installation on-site will be fast and inexpensive. It's anticipated that such homes could be printed in less than an hour and would be inexpensive. Perhaps the most exciting aspect of these buildings is that they're planned to be completely recyclable. The completed homes can be demolished, ground to a certain particulate size, and then combined with a glue-like substance and used again as feedstock for the 3D printer. Essentially, these are disposable houses.[24] The availability of disposable buildings is a good news/bad news situation, depending on what behaviors that phenomenon triggers. Will this cutting-edge technology lead to more sensible management of resources by ingraining a recyclability into our built environment, or will it lead to other forms of wasteful behavior by inspiring "house of the month" clubs or some such variant?

## BUILDINGS WILL BE SMARTER AND MORE AUTOMATED

Smart buildings will emerge during the coming decades. There are many definitions of a smart building, but generally the term applies to buildings with increased sensors, more intuitive automation, better connectivity to the electric grid and other structures, and various levels of computerized management systems for facility operations, with an emphasis on actionable information.

There are many varieties of sensors, levels of automation, and control systems being planned and discussed for smart buildings. Most of the systems focus on energy efficiency, security, remote management, and comfort. They include energy management systems, water management systems, occupancy sensors, facility maintenance systems, asset management, life safety, telecommunications, systems integration, various alarm monitoring and fire protection systems, network intrusion detection and controls, uninterrupted power supplies, surveillance systems, data analytics… the list goes on and on.

The smart building movement in the near-term will focus on home energy management systems, allowing demand response and energy efficiency savings, more advanced security systems, and a few other areas of comfort and security. Products such as Nest thermostats and Amazon Cloud Cam security cameras are part of this trend.

Home area networks and home energy management systems will continue to improve and provide higher levels of control with minimum input. Just as the Nest thermostat learns the user's habits, other energy management systems will also develop rudimentary learning capabilities, making the personal and customized control of the building's energy consumption and generation easier and faster.

Smart appliances will proliferate broadly. Sensors and communication capacity will increase and there will be more whole-house integration of the sensory and control mechanisms. However, integrating all of this into a smart building, and widespread construction and renovation of smart buildings will be neither quick nor inexpensive.

There are several obstacles to our buildings becoming smart. It is much easier to do the design, wiring, and integration of devices in a new building as opposed to retrofitting the cables and communication systems in existing buildings. That means the smart revolution will initially be a phenomenon tied to newer construction. Thus, the spread of smart buildings and interconnected appliances will probably not move much faster than the turnover rate for new building construction and new appliances.

While the number of devices that can be controlled through the internet and a smart device such as Amazon's Echo is increasing, it has not proved easy to get all those things working together. Most people would probably not want to invest the time and money it would take to fully integrate all their different appliances and systems.

Lucid's Michael Murray writes that although getting things like HVAC systems, submeters, lighting controls, and other devices communicating with one another and the internet seems easy enough, that is far from the case. "It's so difficult," Murray says, "that it supports an entire industry called systems integration."[25]

We expect the smart building movement to advance primarily with new buildings while expanding slowly to accommodate widespread retrofits. For the existing infrastructure, there will be single products like smart meters dispersed through all the buildings, but retrofitting the current building stock one product at a time, and then establishing interoperability and integration of multiple products, is a slow process. So we will see a growing disparity between smart buildings and dumb buildings. Expanding smart networks for existing buildings will require wireless networks to avoid remodeling costs. Things like smart meters and wireless communication can be added to older structures, but they will not be able to take advantage of the advanced technology that we will see in Chapter 4.

Building autonomy will also emerge. Certain processes are already being given some autonomy, such as programmable thermostats that are now starting to "learn" our behavior patterns. Automatic doors, faucets, and toilets are just the beginning of building functions that may be handled through sensors and automated services.

The vision of an idealized smart building can be quite grand. Siemens, the giant manufacturing firm, has a promotional video that describes smart building characteristics as "...solutions that turn buildings into living organisms: networked, intelligent, sensitive, and adaptable."[26] We think such a "living organism" is not going to be widespread in the next few decades, but is certainly part of the advanced technology scenario we will explore later.

## CONVERGENCE

The other consequence of the Energy Efficiency Megatrend is the convergence of the building, transportation, and power grids. We are seeing such convergence through on-site generation of power and transportation connections.

### *Convergence with the Power Grid*

At the beginning of the power industry, all generation was on-site. Homes and yachts of the wealthy were the first to be wired for electricity, and they used on-site generators. The first commercial buildings to have electricity were served by generators in the buildings themselves. After alternating current (AC) power won out over direct current (DC) power, Samuel Insull established a new business model, and centralized power plants became the norm.

Insull's visionary innovation recognized that building larger power plants with shared expenses by all the consumers would be cheaper and more reliable than each user having their own expensive, on-site generator. This model worked for more than a century and achieved remarkable reliability and access to electricity.

Distributed generation (DG) is again expanding, and centralized generation is in slight retreat. The primary source for new growth in distributed generation is solar photovoltaic panels. However, on-site generation also includes diesel generators, natural gas microturbines, fuel cells, propane generators, and a few other fuels and technologies.

The Department of Energy's projection for distributed generation to 2040 shows rapid growth of solar photovoltaic, and some growth for the other most common options for distributed, on-site generation:

### *Solar*

Three factors drive the adoption of solar PV: the insolation, or amount of sunlight available; the installed cost of the solar; and the price of electricity from the grid. Many parts of the country are now at or approaching grid parity, where the price of electricity coming from the solar array is equal to or less than the price of electricity from the grid. In those areas where solar panels are a cost-effective option, its adoption occurs relatively quickly.

In regions with less sunlight, or where electric rates are lower, their market adoption rate is slower. DOE assumes penetration of DG as "a function of the estimated rate of return relative to purchased electricity." However, customers also buy solar panels for other motivations, such as their resiliency benefits and environmental merits.

Another reason solar PV may see rapid growth is that the adoption rate for solar PV might more closely resemble consumer electronics than conventional power generation. Traditional power plants that are custom-built on-site, one at a time, must go through permitting, siting, long-term construction for unique large-scale systems, fuel supply acquisition and delivery, and have significant operational requirements.

Rooftop solar panels, on the other hand, are manufactured in factories in large uniform quantities, causing prices to drop and performance to rise. These panels are then shipped through conventional transportation systems and quickly installed by local vendors or building occupants. They do not require the drawn-out permitting processes.

Overall, this sequence offers greater potential for rapid mass deployment than conventional generation sources. Consumers might consider solar on their roofs as being more in the category of a new smartphone or TV than something akin to a power plant. This is especially true as local hardware and furniture stores start to sell and install panels.[27]

Finally, government policy may move from incentives to mandates for solar PV. California has become the first state to mandate solar installations on new homes.[28] Other states will probably follow. Installation of rooftop solar will be severely impacted by the coronavirus pandemic. Access to homes and businesses was generally halted in March 2020 for several months. PV manufacturing in China was temporarily suspended, and thousands of installers around the world were furloughed or laid off. Installations and the supply chain will resume, and this trend remains intact, but the robust projected growth in rooftop PV for 2020 will certainly not be met.

## Natural Gas

Natural gas is also gaining market share of distributed generation through gas microturbines, fuel cells, and conventional natural gas-fired combined heat and power (CHP).

Combined heat and power has great potential for expansion. CHP is not a single technology, but rather several technologies that concurrently produce electricity or mechanical power and also supply useful thermal energy such as steam. CHP systems can operate with up to 80 percent efficiency.

Utilizing waste heat is by far the most economical and productive application of CHP, and it works well in cold climates or near factories where heat is a desired commodity. New York City has more than 100 miles of steam piping underground, serving more than 1,500 buildings.[29] Denmark also has an extensive district heating system, which has the added benefit of providing seasonal thermal storage.

Waste heat is abundant in the utility sector, so finding productive purposes for that heat can be a fruitful path for improving energy efficiency. "The energy lost in the United States from wasted heat in the utility sector is greater than the total energy use of Japan," according to an Oak Ridge National Laboratory study.[30] The study has proposed that CHP could be expanded to account for 20 percent of U.S. electricity capacity by 2030.[31] Unfortunately, there are numerous market and technical barriers to achieving the high potential of CHP.

For example, waste heat does not always achieve the high temperatures necessary for steam turbines, and may be useful only if located near a facility that can use the low-grade heat available. Steam can only be moved a few dozen miles from the source to the customer, beyond which it becomes uneconomical. By comparison, electricity can be moved hundreds of miles with only minimal losses, which is one reason the nation built out a continental electric grid network rather than a steam network.

Gas microturbines represent another aspect of natural gas penetration into the distributed generation market. For years, the price of natural gas was a deterrent to the expansion of these small turbines in the marketplace. With the onset of low natural gas prices in the 2010s, the use of microturbines is expected to expand, but still remain a distant second to solar.

Fuel cells are also primarily powered by natural gas (which is reformed into hydrogen) and are expected to take market share in niche markets that have high reliability and power quality requirements. Japan is pushing hydrogen as a national priority, and companies like ENGIE in France are investing aggressively in hydrogen technologies. These prominent backers, along with significant investments, might push fuel cell adoption at a faster clip.

Though mostly cost prohibitive at present, fuel cells are relatively efficient compared with steam turbines or internal combustion engines. They typically operate with efficiency between 40 to 60%. Some are even 85 percent efficient when waste heat is captured for later use. They are also less polluting, as their reactions occur at low temperatures, thus significantly avoiding the formation of nitrogen oxides. Eventually, fuel cells operating on gas or hydrogen from water resources could be a significant source of very clean power.

Buildings are also establishing closer communication with the power system through smart meters placed at homes and businesses. These meters are a platform for future services and rates from the utility, and can serve as valuable data sources for the building owner and power system.

Energy Management Systems are emerging as the primary control device in a building. Right now they are mainly turning on and off appliances, but they will soon have the capability to respond to utility signals for reducing electrical loads, controlling solar panels, managing filling and discharging batteries for energy storage, and managing when the electric car is charged. All this is part of the integration of buildings with the power grid.

### Convergence with the Transportation Sector

Buildings are also converging with the transportation sector. First, more buildings are becoming fueling stations for electric cars and trucks. Charging stations in garages and other parking spaces provide drivers a convenient way to power up their cars, and many owners are just using the ordinary electric sockets in their homes. At a more integrated level, solar carports allow owners to use their own generated power to charge their cars.

Second, we are just starting to see a trend for cars to power homes when there is an electrical outage. Nissan has offered customers a connecting cord to their vehicle that will light up the home from the car battery. Via Motors has a plug-in hybrid pickup that utility companies can use to power several homes while repairs are being made to the electric lines.

A final way that buildings are merging with transportation is the multiple types of small buildings that are adding wheels: recreational vehicles, mobile clinics, mobile libraries, food trucks, mobile construction offices, and many more.

## CHAPTER 2 SUMMARY

1. Buildings will be much more energy efficient in the future, and the use of robots, 3D printers, and artificial intelligence for design and engineering will lead to remarkably fast construction times.

2. Buildings will become smarter through sensors and the Internet of Things, and whole-house management systems will emerge. Interoperability will be the major obstacle as multiple companies and technical platforms try to work together.

3. Buildings will become more self-sufficient for power through solar and other on-site generation technology.

*The smart sensor built into Bob's pillow began to buzz, signaling that it was time to get his hungover butt out of bed. Thank God for the self-driving car that had carried him home.*

*The heads-up display projected the weather forecast and news headlines on his kitchen window as his coffee brewed. As he reached for the coffeepot, the refrigerator chirped and reminded him he was low on cream.*

*Toast, scrambled eggs, bacon, the usual. The speaker in his bathroom scale reminded him to watch his cholesterol. Texts, emails, and phone call reminders popped up one after another in a corner of the mirror as he shaved. He dictated a reminder to start planning his sister's rehearsal dinner ASAP. He did not enjoy having responsibility for the seemingly endless list of details and logistics. The house made a verbal confirmation: reminder registered.*

*Bob brushed his teeth and swallowed two aspirin. At least he had remembered to plug his car in the night before. Hungover AND late to work was no way to start the day.*

# SUSTAINABLE BUILDING

In this chapter we look specifically at sustainable building technology in the future. Many of the trends highlighted in the previous chapter represent movements toward sustainable building, and this chapter discusses some of the more advanced sustainable technologies and shows how clean and efficient our built infrastructure can be.

Energy sustainability in buildings is achieved through two primary paths: built-in energy efficiency and by sustainability in the external source of the building's power.

Green building technology, when utilized in a building's construction and operation, can help that building realize a sustainable future. Examples of green technology might include advanced construction, low-impact materials, and integrative design and ratings systems.

The external source of a building's power also affects sustainability. A clean and nondepleting power source such as wind or solar in place of, for example, a coal-fired source can yield dramatic results in energy sustainability.

We should also note that energy efficiency is less necessary for sustainability in a fully renewable energy situation. If nonrenewable fossil fuels are used, then energy efficiency achieved by minimizing consumption becomes the highest priority, as it has a direct and important impact on reducing harmful emissions into the air, land or water.

However, if a building is powered 100 percent by a clean, nondepleting, renewable energy source, energy efficiency plays a smaller role in making the entire system sustainable from an environmental impact standpoint.

But as a practical matter, when energy efficiency is prioritized from the beginning, it becomes much easier and more affordable to meet the building's power requirements through the use of renewable energy, so efficiency remains the first step.

## *Energy Efficient World*

In their 2012 World Energy Outlook, the International Energy Agency outlined their Efficient World Scenario (EWS), which explicitly targets energy efficiency potential in future energy infrastructure. Related to this is their 2D scenario—two degrees Centigrade of global warming—which is one of the three scenarios designed to project our energy future in relation to climate change. The two-degree scenario is seen as the necessary energy path to avoid the consequences of catastrophic climate change, so it is consistent with a sustainable future scenario. (The IEA considers the current energy path to be on track to generate about four degrees of warming, and the worst-case scenario calls for six degrees of warming.)

Advanced energy efficiency is a key element in a sustainable energy future. Gains in energy efficiency feature heavily in all the major sustainable scenarios being advanced in policy discussions on climate change.

Buildings are the key element of the energy efficient potential that planners envision. The several sustainable building programs in place in the U.S. and Europe are starting to have a real impact on new construction. Zero energy buildings are increasingly touted, and even mandated, in some building codes.

Another key element of the sustainable building scenario is the power production potential of built-out infrastructure. This consideration encompasses the proliferation of clean distributed energy generation sources and the eventual evolution of local microgrids that combine efficiency, distributed generation, and energy storage. These microgrids will interconnect seamlessly with the larger grid. Both building and power innovations will need to happen in parallel to maximize a fully integrated and sustainable built environment.

The IEA's Efficient World Scenario assumes a high level of installed efficiency worldwide in the next two decades. This scenario depends on buildings to achieve 41 percent of the desired savings by 2035.[1] The potential is significant. New buildings could use less than 10 percent of the energy in current designs, and existing buildings could be retrofitted to reduce their thermal energy needs by 90 percent. Equipment and appliances could be 5 to 90 percent more efficient than current designs.[2]

Space heating and cooling is and will remain a priority in many developing countries because of changing climate conditions and demographic trends toward higher affluence. As people get richer, their energy needs expand to include a comfortable environment in their buildings. The IEA notes that when the 50 largest metropolitan regions in the world are considered, the vast majority are in hot climates and in developing nations, and in need of air conditioning.[3] It is assumed that efficiency gains in modern technology, such as air conditioners, will gradually become the global norm.

Due to the continued use of wood for fuel in these countries, cooking also presents a key opportunity for energy savings. Advanced fuels and cook stoves could replace wood-fueled cooking. The accompanying slowdown in deforestation could also have a salutary effect on the environment.

Lighting and appliances are identified as having the greatest energy savings potential in the near future. Additional savings are assumed in this scenario by better design techniques for buildings and more automated controls, which can optimize the operation of appliances and building systems.[4]

The EWS model makes a major assumption that applying energy performance standards to 95 percent of the equipment in all countries by 2035 will lead to greater efficiency levels in appliances across the board.[5] As more countries adopt performance codes for more appliances, energy efficiency will gain more traction. A refrigerator sold in India will ideally have the same energy efficiency as one sold in Kansas.

The analysis of the U.S. market from the IEA shows a potential reduction in residential demand to below the current energy demand, achieved largely through insulation and space heating. After efficiency, IEA expects the next-biggest savings to come from technology and fuel switching, such as shifting from gas-fired boilers to electrically driven heat pumps.

The IEA then notes a major difference in the manner by which the efficiency targets of the Organisation for Economic Co-operation and Development (OECD) member countries and non-OECD countries will be achieved. In general, these can be viewed respectively as developed and developing countries. In developed countries, where new construction might be as low as 1 percent of the building stock each year, efficiency gains will be made primarily in more efficient retrofits of existing buildings. An example of such a retrofit is the reduction of space heating and cooling requirements via improved insulation of existing building shells. In the U.S. and Europe, a significant portion of building-related energy savings will come from improving the insulation within buildings that exist today and that will still be standing in 2035.

In developing countries, where construction is quickly expanding, the biggest savings potential is in new buildings, and their use of new and more efficient appliances and equipment.[6]

It is ironic and sad that the popular "Green Building" movement, and the push for more efficient building codes, is happening chiefly in the U.S. and Europe, where there is and will be little new construction compared with developing countries. In the latter, new construction will occur on a relatively large scale, but most of the work will be done without efficient energy codes, or without any building codes whatsoever.

The International Energy Agency cites building codes as fundamental to achieving a sustainable building scenario. They state that "progressively more stringent building energy codes and minimum energy performance requirements for all significant energy-using equipment is the key policy for the buildings sector."[7]

## SUSTAINABLE BUILDING

How do we achieve sustainable building, and what does that mean? A good reference point is to look at the rating systems for sustainability in new construction.

### *Rating Systems, Codes, and Standards*

How, when, and where building codes and standards are implemented will play a key role in achieving a sustainable building future. There are several national, state, and local codes and standards for sustainable or green building, but the most well-known and referenced standards are the LEED rating system of the U.S. Green Building Council, the American Institute of Architects 2030 Commitment, and the Living Building Challenge.

LEED stands for Leadership in Energy and Environmental Design, and was the first national rating system and standard for sustainable buildings in the U.S. The U.S. Green Building Council developed LEED in the early 1990s, and its rating system, which ranges from simply certified to top-tier platinum, remains the standard for sustainable building planning for many cities and organizations. The certification scorecard covers sustainability in water and energy efficiency, air quality, materials and resources, indoor environmental quality, innovation in design, and regional priorities.

Another major sustainable building initiative is the 2030 Challenge. Formed by Ed Manzia and his nonprofit organization, Architecture 2030, the 2030 Challenge is a push for all new buildings to be carbon-neutral by 2030. The ZERO Code defines a zero net carbon building as a highly energy-efficient building that produces on-site, or otherwise procures, enough carbon-free renewable energy to meet building operations energy consumption annually.[8] To qualify, a building must reach the goal by sustainable design, on-site renewable power, and purchasing up to 20 percent renewable energy.[9]

The American Institute of Architects, the U.S. Conference of Mayors, and the National Governors Association have accepted the challenge. The Energy Independence and Security Act of 2007 mandated that all federal buildings be carbon-neutral by 2030. While federal buildings are a small slice of all the buildings in the nation, it is nevertheless significant enough to help foster the construction industry's capabilities and supply chain for adoption elsewhere. The AIA challenges member architects to design buildings to reduce the estimated 40 percent fossil fuel consumption of buildings today to carbon-neutral. A 2013 survey by *Design Intelligence* found that 52 percent of U.S. firms had adopted the 2030 Challenge.[10]

A third sustainable building movement currently gathering momentum is the Living Building Challenge. This rating system goes beyond the areas in the LEED rating system to include beauty and inspiration. The rating isn't awarded until actual performance is measured after a year of operation.[11]

Seattle's Bullitt Center is the first speculative development to achieve Living Building Challenge certification. It utilizes daylight as its primary illumination source, and also incorporates advanced geothermal and heat recovery systems, elevators that generate energy while they descend, rainwater-based water systems, composting toilets, and an automated blind system to optimize temperature control. This remarkable building performs at 79 percent below Seattle's already stringent 2009 energy code.[12]

## WHAT IS THE POTENTIAL?

As substantial as the efficiency gains assumed in the IEA's Efficient World Scenario may be, Amory Lovins and the team at Rocky Mountain Institute see even greater potential reduction in future demand. In their book *Reinventing Fire,* they say that there could be a 69 percent savings from the business-as-usual projections of the Department of Energy. This potential reduction could be achieved through a combination of efficient technologies, smart controls, and integrative design.[13]

The EIA does not consider mass deployment of the advanced efficient products that Lovins does. The EIA projection is probably a more realistic view of energy consumption in the near future, but Lovins demonstrates the potential of these technologies to meet our needs with less energy conversion. Some of these technologies will see exponential growth beyond IEA expectations.

The technologies that Lovins highlights demonstrate the major elements of the Energy Efficiency Megatrend: reduction of materials, motion, and time by increasing the information intensity of the technologies.

Material reduction will result from the recycling and use of new materials that could meet energy needs with less mass (such as better insulation). Time and motion will be saved through on-site generation, reducing the need for fuel conversion and subsequent transmission and distribution to sites.

Embedded information shows up in increasingly sophisticated insulation, phase-changing materials, smart windows, roofing materials, solar power generating roofing tiles, and much more. Smart appliances control power consumption and enable intelligent efficiency by controlling the timing of energy consumption. Ubiquitous sensors that give detailed feedback on environmental and operational conditions allow the building to consume power only when it is needed, and to generate and consume, store, or dispatch that power when appropriate.

## HOW DO YOU MAKE BUILDINGS ENERGY EFFICIENT?

Energy efficiency in buildings results from three main components: building envelopes, appliances, and building operations. A building's envelope, or shell, is the primary determinant of how easy it is to heat and cool a building. The better the insulation and windows, the less energy it requires to heat and cool the building. The efficiency of a building's appliances also plays a key role in its overall energy consumption, as does the operation of those appliances.

The potential energy savings from these three components come in the same order, with the most dramatic savings achieved by changes in a building's envelope, followed by efficient appliances, and finally, smart controls to more efficiently guide a building's operation.

### Building Envelopes

The greatest gains in increasing building efficiency have occurred for the largest loads: heating and cooling. We can expect future gains in efficiency in the same areas, primarily through advances in the performance of the building envelope—walls, roofs, windows, and insulation. Passive solar design can also contribute greatly to these efficiency increases.

In *Reinventing Fire,* Amory Lovins and colleagues detail many advances in building technology that play key roles in his sustainable vision. Several of the technologies he cites reduce space heating and cooling, and thus could have the largest effect on overall energy load reduction. Some, like insulation, smart windows, and phase-changing materials, reduce heating and cooling by embedding information in the building envelope. Others, like heat pumps and LED lighting, reduce consumption by increasing appliance efficiency through design and engineering.

There are several sustainable options for new coverings for buildings other than traditional steel, windows, sidings, etc. Choosing the most sustainable option for the building skin will depend on both local climate conditions and the purpose the owner wants to emphasize. The outside surface could be used to help power the building through solar technology, or organic coverings could be selected for both aesthetic and air-filtering purposes, and windows of different sorts might be chosen for viewing, climate control, or power. Because most building skins can do only one thing at a time in a given space, the best skin for a building's shell depends on the purpose of the building, or specific part of the building.

### Insulation

A building material with great potential for gains in performance is insulation—or what Lovins calls radical insulation. Fueled by breakthroughs in nanopore technology and aerogels, radical insulation is substantially better than traditional insulation. Nanopore insulation contains nanoporous silica and carbon particles that create a material with pore sizes less than a hundredth the size of those in fiberglass, polystyrene, or polyurethane. Because of the tiny pores, gas molecules within the material are more likely to collide with the pore walls than with other gas particles (a process known as Knudsen diffusion), resulting in very little gas energy exchange, which, in turn, gives nanopore insulation its tremendous thermal performance. NanoPore, Inc. claims their product, Thermal Insulation, with pores only 10–100 nanometers wide, can possess thermal resistance values seven to eight times greater than conventional foam insulation materials.[14]

Aerogels are another emerging technology that offers significantly better insulation with less material. Aerogels are created when synthetic gels are dried in such a fashion that the liquid component of the gel is replaced with a gas, leaving the porous, solid matrix of the gel intact. What remains is a super low-density material with pores ranging from 2 to 50 nanometers in diameter. Consequently, silica aerogels exhibit the lowest thermal conductivity of any known solid, while also being among the lightest materials on the planet.[15]

A third improvement in insulation will come in the form of phase-change materials. This insulation works by absorbing and releasing heat to balance temperatures. These materials, such as sodium hydrates, change from liquid to solid and back again as the temperature changes. They absorb excess heat, changing from a liquid to a solid, and release heat if the temperature goes below a given target, changing back to a liquid. This absorption and release of heat is significantly more efficient than solely using HVAC equipment to heat and cool the air to maintain a target range. BASF Corporation makes a phase-change material named Micronal PCM, which is a microencapsulated, high-purity paraffin wax that changes phase at 73°F. National Gypsum Company is testing a panel called ThermalCORE using Micronal to absorb heat during the day and release it in the evening.[16]

### Heating and Cooling

Just as motion sensors turn lights on and off in response to our movements through the house, similar sensors now cool or heat rooms in much the same manner. In the future, we may target even smaller spaces within a particular room, and eventually create systems so finely tuned as to heat or cool a single person to their preferred temperature in only the space they occupy. Rather than heating or cooling an entire house to achieve a comfortable climate for sleeping, what if we had mattresses that cooled or heated just the sleeping space to the desired temperature?

On a larger scale, district heating and district cooling could see further expansion in urban and commercial areas. The city of Austin, Texas has a downtown district chilled-water facility that saves commercial building owners money by chilling water off-peak on

summer nights and pumping the chilled water through those buildings during the afternoons, significantly reducing cooling loads. This
approach also saves building owners a considerable amount of space
that otherwise would have been required for on-site cooling systems.

The city of Minneapolis, Minnesota has operated both district
heating and district cooling systems for decades, and delivers heat
from boilers to approximately 100 buildings, and chilled water to
more than 50 buildings in the downtown area. The potential for
substantial reductions in heating and cooling loads by expanding
district heating and cooling to other areas is immense.[17]

### Solar Cooling Paint and Panels

A new type of solar paint could help cool homes. The paint, developed
by the Israel-based start-up SolCold, absorbs heat from the sun and
re-emits it as higher-frequency light, producing a net cooling effect.
According to simulations, applying the paint to the roof of a house
could make a top-floor room feel up to 10°C cooler. Lowering the
temperature inside a room sets SolCold apart from existing cooling
paints, which only reduce the heat absorbed by buildings. Shopping
malls and stadiums are expected to be among the company's first
customers.[18] A start-up in the San Francisco area named SkyCool
Systems does something similar, but with panels. Instead of traditional
rooftop panels that convert solar radiation into heat or electricity,
their panels convert heat into blackbody radiation at a particular
wavelength that will not be absorbed by atmospheric water vapor.
By using radiation to reject heat to space, the panels and paint work
around the clock and give steady, passive cooling.[19]

### Better Windows

Solving the problem of letting in light but keeping out heat has been a
technical pursuit for decades to reduce cooling loads in hot climates.
Various solutions have emerged, including Lawrence Berkeley National
Laboratory's "superwindow," which adds an ultra-thin layer of glass
to double-glazed windows and replaces argon with better-insulating
krypton gas to significantly improve efficiency. The lab is commercial-
izing the product with window manufacturers. Products with higher
efficiency are constantly in development. Designs that use the same

form factor as existing windows simplify the task of swapping out the windows as expensive renovation costs can be avoided.[20]

## Smart Windows

Smart windows control their transparency, and thus the amount of heat and light transferred between the environment and a building, by electrical, photonic, or chemical means. While there may be many types of smart windows in the future, there are at least four types of smart windows under immediate development today: electrochromic, liquid crystal, thermochromic, and photochromic. These windows can be tuned to allow different portions of the light spectrum through, with different trade-offs in terms of illumination, heating, and cooling.

Electrochromic smart windows change their opacity in response to current or voltage, changing from transparent (when maximum light and heat transfer into a building is preferable), to translucent, when minimum light and heat transfer is more desirable.[21]

Liquid crystal smart windows use a different technology. In polymer-dispersed liquid crystal windows (PDLC), liquid crystals are dissolved in a solid polymer, and then sandwiched with a thin, transparent conductive material between two layers of glass or plastic. Like electrochromic windows, PDLC smart windows react to a charge, which causes the liquid crystals to align with each other, allowing more light, and thus more heat, through the window.[22] One company, SONTE, has developed a tinting treatment using this principle. The window film is like an ultrathin LCD TV and a corner of the film plugs into an electrical transformer that is enabled for wireless operation so the owner can control the panel through an app or a switch. SONTE plans to upgrade the system to connect to smart house controls that will turn it on and off relative to the temperature.[23]

Thermochromic windows work in yet another fashion. They change their color and opacity in response to changes in temperature. Most often, thermochromic windows are designed to become more opaque as heat increases and more transparent when it's colder. Unfortunately, they can become nontransparent and unsuitable for view windows.[24]

Finally, photochromic smart windows act like the popular sunglass lenses that darken during brighter conditions and lighten when it's darker. Although this application is effective in reducing heat transmission in bright summer conditions, their effectiveness is limited in winter during bright conditions, when they still darken in direct sunlight.[25]

## Solar Windows

How about actually generating electricity from the windows? Energy-efficient windows have a coating that blocks ultraviolet and infrared photons. Photovoltaic windows absorb those photons and, unlike the above-mentioned smart windows, actually generate electricity.

Solar Window Technologies uses organic photovoltaics (OPV), and the company licensed its spray-on coating process to Triview Glass Industries in 2017. Transparency, color, and tint determine the power output of the glass, but a 50 percent transparent window could produce 50 watts of energy per square meter, or enough to power eight cell phone chargers.

However, stability and degradation are problems with OPV technology. The company UbiQD is trying to solve this problem using a semiconducting nanocrystal called a quantum dot. Quantum dots can fluoresce and be embedded in transparent material within a solar window to form a luminescent solar concentrator. Trapped light can then be redirected to a nontransparent solar cell on one edge of the window. The companies Glass to Power in Italy and Physee in the Netherlands both use this technology.[26]

The largest PV window installed as of late 2019 was a skylight in a London office building that was produced by the Spanish company Onyx Solar. Onyx Solar uses lasers to remove the opaque silicon layer and back contact from thin-film solar panels to make them more transparent. The company also has installations in the NBA's Miami HEAT American Airlines Arena in downtown Miami, and agreed to install the world's largest PV skylight in the Bell Works building in New Jersey.[27]

## Sustainable Wood Buildings

Along with these impressive emerging technologies, we will also gain efficiency through using sustainable and recycled materials in new and advanced ways. For example, consider the plyscraper movement. Engineers and architects have now developed innovative methods for constructing skyscrapers made almost entirely of timber. Using mass timber or cross-laminated timber (CLT)—essentially, boards of lumber bound together with strong structural adhesives—mass timber has a high strength-to-weight ratio while being up to six times lighter than concrete.[28]

Wood is a good building material because it is lightweight, strong, and earthquake resilient. The thick, treated pieces used in the method are also fire-resistant, forming a char on the outside layer when burned, protecting the underlying material. Such structures could soar far beyond the four or so stories presently limited by many areas' building codes, and could save energy and materials as well as significantly reducing carbon emissions. Though such wooden towers would use a good deal of timber, managed forests offer a sustainable source of building materials. The Programme for the Endorsement of Forest Certification lists more than 424 million hectares of managed forest through 20,000 companies.[29] Mass timber also lends itself to shorter building times and less construction traffic, and thus fewer greenhouse gas emissions during construction.[30]

A 10-story apartment complex was built in Melbourne, Australia and a 14-story timber apartment building was constructed in Norway.[31] The world's tallest timber-based building, Brock Commons, opened in Vancouver at 18 stories.[32]

The architectural firm Skidmore, Owings & Merrill calculated that a 125-meter tall skyscraper made mainly of wood would be a feasible project. Along with being about half the weight of steel and concrete buildings, timber and concrete structures would produce less than a quarter of the carbon emissions as conventional construction.[33]

According to International Code Council's (ICC) Andrew Tsay Jacobs, code acceptance will be the strongest driver of mainstream use of mass timber.[34] In the fall of 2018, after an ICC hearing, 14 code proposals concerning mass timber were recommended for approval by the greater ICC community.[35] With Europe and Canada leading the pack, other countries, including Australia and the U.S., have begun mass timber construction projects, with even more projects pending.[36]

## APPLIANCE EFFICIENCY

### Air Conditioning

Air conditioning has the potential for widespread and impressive efficiency gains. One innovation from the National Renewable Energy Lab (NREL) is the desiccant-enhanced evaporative cooling system (DEVap). In understanding the advance that this machine represents over traditional air conditioners, it's important to remember that air conditioners control inside air by removing both heat and moisture. Both processes—dehumidification and cooling—require energy. The DEVap cooling system developed by NREL thus has two stages; first removing moisture by use of a liquid desiccant, and then cooling the air with an indirect evaporative cooler.

This DEVap arrangement has been shown in models to dramatically reduce air conditioning load, and can reduce energy consumption by 30 to 90 percent as compared with traditional air conditioners, depending on the climate and humidity. Furthermore, they do not use the harmful refrigerants used by traditional air conditioners.[37] DEVap systems are already in development by many leading air conditioning manufacturers. They could have a significant impact on the efficiency of the overall built infrastructure, and are a testament to modern appliance efficiency through improvements in design and engineering.

### Heat Pumps

Heat pumps work by moving heat from one location to another, essentially like an air conditioner but in reverse. They require a source for heat, which can be a water body, the outdoor air, or the ground. The most common and cheapest versions are air-source and ground-source heat pumps. It's more efficient to pump heat from those sources into the building rather than directly heating or mechanically cooling the indoor air. Systems that are integrated with water heaters improve efficiency further.

Ground-source heat pumps are appealing since the temperature below the surface remains fairly constant throughout the seasons. Another advantage of ground-source heat pumps is that they rely on a globally available resource, though in some locations it can be expensive to bore the wells to reach the right depths to make the process efficient. However, for really cold climates, heat pumps lose their effectiveness. Using advanced designs, some of the common off-the-shelf versions available in 2020 work with outdoor air temperatures as low as –5 to –15 Fahrenheit but at temperatures below that threshold would require supplemental heating from electric devices or combustion.

### Advanced Lighting Efficiency

The revolution underway in lighting technology could offer the quickest gains in building efficiency. There has been a dramatic increase in the efficiency gained as we moved from the incandescent light bulb to compact fluorescent lighting to light-emitting diodes.[38]

The next advance in lighting efficiency is expected to come from organic LEDs (OLED). In OLEDs, a conductive and emissive layer of organic molecules or polymers is inserted between a cathode and anode structure. When subject to an electric charge, the organic materials within the OLED emit light. This technology is rapidly being developed for televisions, display screens, and lighting applications. Displays will be lighter, thinner, brighter, and more efficient than LED displays, and OLED lights will become the most energy-efficient lighting available.

## BUILDING OPERATIONS

### Intelligent Efficiency

The American Council for an Energy Efficient Economy (ACEEE) has labeled the energy efficiency gains from smart grid devices as intelligent efficiency. This refers to energy efficiency gained from sensors and processors built into buildings to promote the automation of appliance and energy usage. Motion-controlled room lighting and advanced thermostats that "learn" your temperature preferences and lifestyle habits, available today, are examples of using information and communications technology to reduce appliance consumption.

A 2012 report by ACEEE estimated that the U.S. could cut energy consumption 12 to 22 percent by moving to a greater degree of intelligent efficiency. This is an improvement from component efficiency, where energy efficiency improvements focus on individual appliances. The intelligent efficiency approach takes advantage of new information and communication technologies to optimize the performance of the system as a whole.[39]

Much is made today of the deepening Internet of Things, namely the connections between devices and the internet for purposes of information transfer, interpretation, and, perhaps, automated redirection of device usage. Household appliances such as refrigerators and televisions are already incorporating online interconnectivity.

Of course, the more automated a building becomes, the more dangerous a power outage becomes. Not just elevators, but windows, doors, and numerous other aspects of buildings could lock down, necessitating manual overrides for safety.

### Integrative Design

The efficiency gains within a building can be raised to a much higher level by the use of integrative design. Whereas efficiency gains can be achieved piecemeal by upgrading lighting, heating, or cooling systems, a smarter way to achieve even greater overall gains would come from designing systems to work more efficiently as a whole.

An excellent example of this is the design for a typical water pumping and plumbing system. The following chart illustrates the energy efficiency for typical plumbing systems, showing how an energy output of 100 units at the power plant is wasted at each stage of the process until only 9.5 units arrive at the end of the pipe. Energy losses are compounded as the liquid moves through the system. The chart also implies how improving efficiency at each stage in reverse could compound the savings in the same fashion. An example of such a change is simple piping redesign, replacing right-angle plumbing with a geometry that reduces friction. Incremental changes made throughout the system could result in a substantially more efficient plumbing system.

# ENERGY EFFICIENCY FOR TYPICAL PUMPING SYSTEMS

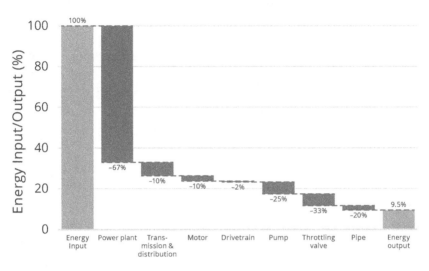

Source: Amory Lovins, *Reinventing Fire: Bold Business Solutions for the New Energy Era*

**Figure 3.** Full energy efficiency for pumping systems

## RENEWABLE ENERGY

To this point, we've been discussing the first component of a sustainable building: energy efficiency in the construction and operation of its structure. The second component of a sustainable building is its energy source. To be truly sustainable, a building must be powered by a clean, renewable energy source. Grid-sourced energy offers economies of scale but has a sprawling reach that can be costly and impactful. While on-site sourcing does reduce transmission and distribution costs and infrastructure somewhat, it misses out on some of the economies of scale while adding material and connectivity to the site itself.

In the previous chapter we noted the exponential growth in solar photovoltaic (PV) power on rooftops. Solar and other on-site renewable energy technologies will be a dominant part of our sustainable landscape. Sunlight is a ubiquitous source of clean energy, and merits further discussion of how best to use the sunlight that strikes a building site. We think the priorities should be passive solar design, followed by active solar thermal water heating, PV power generation, and, finally, green roofs.

## *Passive Solar Design*

Renewable, clean, on-site energy in a building refers to capturing and using some energy flows, like sunlight and wind, while averting other energy flows, such as waste heat. This means that even before installing solar panels or solar water heaters, there should first be consideration of how to best utilize what energy flows the surrounding environment can offer a building.

Thoughtful orientation of the building to both the sun and prevailing winds, and the placement of windows and window overhangs, can often meet a significant part of a building's energy needs without active energy conversions. Passive solar design can be the decisive feature of a sustainable building, making energy loads manageable by active on-site generation. And when sunlight and wind are intelligently captured, and heat averted, loads on active energy systems will be significantly reduced. Appliances that generate waste heat, such as refrigerators, should be situated in a way that the heat is put to work.

## *Active Solar*

After utilizing passive solar design, it then makes sense to consider active solar, whereby sunlight can be converted to both thermal and electric energy. Though often considered a secondary application, solar thermal systems (which refers to solar water heating) directly utilize the solar radiation already hitting your roof without additional energy conversions. A good solar water heater will convert more than 40 percent of the incoming sunlight into heating water. On the other hand, solar photovoltaic panels will produce electricity at a typical conversion rate of less than 25 percent for current solar panels. And then that electricity may be used to heat water, a less efficient process.

While solar PVs are trendy and can provide services other than heat, a solar hot water heater is a more cost-effective, efficient, on-site energy application for a building's roof, and should thus be the priority, followed by covering the remainder of the roof with solar panels for generating electricity.

These priorities only make sense if a roof meets solar require-ments. In some cases, a building's orientation and tree cover could preclude solar. Because tree cover is excellent for energy efficiency, it's unwise to cut trees to make solar more feasible. If solar can't be used on all or a portion of a roof, then it likely makes sense to paint the roof white to reflect sunlight and reduce the heat transferred into the building. If the climate and water availability are suitable, a green roof—one covered with plant life—is an intelligent option and provides many benefits for both the building and the environment.

### Other On-Site Energy Flows

Depending on a number of geographic and climatic factors, low-power wind (using micro wind turbines) could also provide a clean, renewable, on-site energy source for buildings. There are other smaller-scale energy flows, such as temperature differentials, noise, and vibrations that could be captured and redirected. Such sources of energy would be ideal for lighting, sensors, and charging small batteries. Some buildings may contain piezo-electric materials that use the vibrations from a crowd walking on a floor to generate light-ing or other utilizations.

### Matching Fuels with Workloads

There are many of ways to generate electricity and meet energy workloads for a building. Building workloads have different power requirements, thermal needs, and so forth, so a key to the most efficient use of energy within a given building is properly matching available fuels with workloads.

For example, solar is a good choice for meeting many of the low-power electrical loads of a building, such as lighting. Although solar can power a high-power appliance, like a plasma TV, it is not best suited to doing so. Devices such as air conditioners, teakettles, computer servers, and clothes dryers could quickly use up all the solar available on a site, even with added energy storage.

It would be more efficient to use energy created through passive solar design, solar thermal hot water heating, and solar photovoltaic to meet the relatively low power requirements of the building. Solar-plus-storage can meet all the energy needs of a given environment if the energy loads are low enough. However, it may be necessary to meet high-power requirements through grid-connected, renewable-based electricity, using the grid for energy storage when producing more energy on-site than is being consumed.

While properly matching fuels and workloads, like solar-produced energy to lighting, might sound complicated, elegant advances in smart controls are already solving the problem. Advanced home management systems can gather, convert, and dispatch energy on-site to where it is needed.

## ZERO ENERGY BUILDINGS

A standard growing in popularity with the the sustainable buildings movement is the zero energy building, or ZEB. Zero energy building (ZEB) is, at present, a rather amorphous term with several definitions. In a September 2015 report, the Department of Energy defined a ZEB as "an energy efficient building where, on a source energy basis, the actual annual delivered energy is less than or equal to the on-site renewable exported energy."[40]

Almost all ZEBs are technically *net zero energy* buildings, since it's assumed that the electric grid will provide power when needed, absorb excess generation from the building site, and credit it to total use, thus delivering net energy consumption to near zero. Usually, net zero energy buildings are averaged over the course of a year, producing more energy in the summer months and consuming more in the winter months.

To be considered a zero energy building, "a ZEB may only use on-site renewable energy in offsetting the delivered energy. On-site renewable energy is energy produced from renewable energy sources within the site boundary."[41] This means that the building's energy needs must be met by solar, wind, geothermal, water, or other energy flows that impact the site.

Austin, Texas has a city building code that requires all new single-family residences to be zero energy capable homes. This means that they will be so energy efficient that if solar panels are placed on the roof, the net energy consumption of the home over the course of the year will be near zero. Austin's code prioritizes building in an energy-efficient manner before considering renewable energy. This represents the only workable way to get to ZEBs, as building consumption must be sufficiently lean before on-site power generation produces a net ZEB.

As we pointed out in chapter 1, it won't be easy to fully power multifamily housing, commercial office buildings, restaurants, and many other types of buildings by on-site renewable sources of energy alone. The energy density of solar and wind occurring on the footprint of the building itself will not be sufficient to meet the power density requirements of the building. They might be able to convert on-site biomass, such as food waste, into a source for heat and electricity, but the mass of waste may not be sufficient to meet energy demands.

In fact, some present-day ZEBs will only be so until they plug in an electric car or get a 3D printer, because the energy demands of those items will exceed most building's on-site energy sources. While 3D printers and electric vehicles provide other benefits in terms of society's total energy consumption, they complicate the push for on-site energy reductions in the residential and commercial sectors.

That doesn't mean, however, that such structures can't be powered by clean, renewable energy. Wind, solar, biomass, geothermal, and ocean power are all clean, renewable sources of energy that, if built to utility scale, could deliver power via the grid.

This point about high-energy demand is especially important when discussing net zero energy commercial buildings. California, for example, has established goals for net zero energy building codes for residential buildings by 2020, and for commercial buildings by 2030. We are extremely skeptical, for the reasons mentioned above, that true net zero energy could be achieved for most commercial buildings.

Although solar on-site generation will become much more efficient, the conversion efficiency can never exceed 100 percent. No matter how efficient the building envelope and appliances become, it is difficult to see how commercial buildings with substantial loads—either electric or thermal—and a narrow physical profile can ever be completely powered by the energy flows impacting the building site. Multistory buildings with small footprints, as is typical in Manhattan and other big cities, or buildings housing large amounts of computer equipment, or restaurants, or heavy industrial processes, will need additional power sources.

Perhaps, rather than mandating net zero energy codes for commercial buildings, we should mandate very high efficiency standards and fully utilize all on-site generation opportunities. That's why the City of Austin doesn't have a net zero energy commercial standard, but instead is aspiring to set a very high energy efficiency goal as a standard.

In general, it will be impossible for a very tall, skinny building on a small lot with high office equipment loads to be a ZEB, but relatively easy for a long, flat, un-air-conditioned warehouse to reach that goal. Ultimately, we'll see a lot of single-family ZEBs with advanced technology, and at least a portion of the commercial building sector will be ZEBs.

Spacing and solar access codes will also become important. A ZEB, such as the remarkable Bullitt Foundation building in Seattle, could quickly lose that status if the construction of tall buildings on adjoining lots block the sun. The commercial sector can certainly be fully powered by clean, renewable energy, but usually that power will need to be delivered from the electric grid rather than on-site renewable energy. Thus, zero net carbon (ZNC) building codes will be much more readily achievable and practical than mandating ZEBs.

If this scenario plays out, we expect almost the entire residential and commercial built infrastructure will eventually be composed of highly energy efficient and either zero or close-to-zero energy buildings, powered by a combination of on-site renewables and grid-powered, utility-scale renewable energy.

## WILL OUR BUILDINGS BE SUSTAINABLE?

Achieving the Efficient World Scenario will be a stretch. The targets for building efficiency set by the International Energy Agency to limit global temperature rise to two degrees is not being met.[42] The change in codes and standards and deployment of new materials are just not happening fast enough.

The Lovins vision shows us a potential future. These technologies and others like them promise a future building stock that is dramatically more efficient and clean-powered than today. There could be buildings that use only 10 or 20 percent as much energy as a comparable building today.

One of the most exciting developments is solar power becoming cheap and abundant; we see solar on everything. New materials will significantly reduce energy consumption, and smart appliances will at least dampen the large growth in electrical consumption due to consumer electronics.

Unfortunately, these advanced technologies are not going to become the standard installation in new buildings anytime soon. They are usually not going to be cheaper than the alternatives, such as standard insulation or off-the-shelf air conditioners. Government regulations for building codes and appliance standards take years to put in place, and that is only in the cities and countries where the government is concerned about efficiency in energy consumption.

Nevertheless, these building innovations and technological advances will eventually become dominant. Much of the technology being used now for environmental sensing and other information gathering will be crucial to the development of sentient-appearing buildings that interact with us and other elements of the interior environment.

## CHAPTER 3 SUMMARY

1.  Buildings are going to become highly efficient through changes in the building envelope (insulation, windows, building materials), high efficiency appliances, and intelligent building operation and design.

2.  Buildings will become smarter about using on-site energy flows such as solar and heat, and will respond to changes in temperature and light with intelligent materials and appliances.

3.  There will be many zero energy buildings, but most people will live and work in net zero carbon buildings in urban areas.

4.  Sustainable building practices will eventually become standard, but it will take many years to replace current materials, processes, building codes, and construction practices.

"As it fuses more and more with AI, our buildings will become "living" spaces that constantly adapt to meet our needs intelligently and thoughtfully."

— AMIR HUSAIN

*The Sentient Machine*
*The Coming Age of Artificial Intelligence*

Bob lumbered awake and stumbled to the bathroom as the house greeted him far too cheerfully. Still hungover from the night before, he hoped the toilet wouldn't tell the refrigerator not to order any more beer. Fortunately, the toilet light remained green and made no mention of last night's revelry.

The house filled him in on the news of interest and there were no urgent messages, although his sister had sent an inventory of to-dos for her wedding. Bob deftly delegated the list to the house, which he had named Alex to honor his favorite movie star.

As the kitchen served his breakfast, he suddenly remembered the prior night's conversation about attending the Boston meeting.

"Alex, I need to be in Boston on Thursday at 10 to meet with Frank and his staff." As he dropped the dishes into the sink for Alex to clean, travel options appeared on the heads-up display on the touch-sensitive kitchen window, and he lazily chose the optimized cost and environmental option—not the fastest, as he hoped to catch up on sleep during the trip.

As he dressed to get out the door, Alex informed him of household news: the sun room would be transformed into an extra bedroom for his cousin's stay next month; shrimp and wine had been ordered for Friday night's gathering, and budgetary limits were honored; and there was some hail damage from the thunderstorm last night and so repairs were underway by the self-healing roof tiles.

Bob mumbled his thanks as he headed out.

# SENTIENT-APPEARING BUILDINGS

The buildings of the future may seem to have minds of their own. They will supply their own power and repair and reconfigure their structures. They will be able to relocate. They will converse with their occupants and the outside world and, if given the authority, make decisions about their structure, function and power resources. They will handle deliveries, visitors, weather, and much more. These buildings might seem like living organisms at times, responding to both human and environmental changes.

In this chapter, we first look at the elements that will distinguish an advanced technology building: the computational and communication capacity that makes it appear as a sentient being, the materials that give it shape-shifting powers, and its automation. In the process, we talk about the experience of living and working in such an environment. Finally, we discuss some new roles these advanced buildings might play in our lives.

## SENTIENT-APPEARING BUILDINGS

In the Syfy Channel's program *Eureka*, one of the protagonist's homes is governed by a sentient-appearing system named SARAH, for Self Actuated Residential Automated Habitat. As can be expected from a television drama, SARAH developed all sorts of relationship issues with the occupants and town. Today such buildings of the future seem more science and less science fiction.

A building's capacity to control its functions and communicate with people and the outside world is dependent on the computational capacity and communication skills of the artificial intelligence either embedded in the building or linked to it. There will be at least three different levels of intelligence and responsiveness in a building.

First is the interconnectivity of objects that have embedded sensors, the capability to respond to instructions and their environment, and that are connected to the internet. These range from thermostats to appliances that are currently controlled by your smartphone. The building will be filled with what David Rose calls "enchanted objects."[1] Things may respond to your voice or a gesture, or just silently respond to environmental changes such as temperature or light.

The second level of intelligence—let's call it the building manager—could also control these objects. Today we have building automation systems that manage the heating and air conditioning systems, lights, security systems, and more. We see this control system expanding and becoming increasingly intelligent until it manages *all* the house functions, including the structure itself and determining the degree of communication and decision-making regarding the outside world.

This system can become quite sophisticated and interact with occupants in a complex way, almost as if it were a person. It may or may not be physically located in the building; the computer core containing the building management system might be in a server somewhere off-site, while the sensors and actuators and microprocessors would be embedded in or attached to the building structure.

The third level of intelligence in a building is the highest level of AI available through the internet. This is what you experience today when you interact with Alexa and Siri, for example.

This AI of the future could be an impersonal entity, or the building interface could also communicate with your personal digital assistant. That would be a personal AI formed to interact with other AIs as your representative, complete with all your shopping preferences and other personal information. This personal AI would speak to you through the house, but also follow you in the car, over your phone, or even within other public buildings or places where a connection could be made.

Thus, the building will become a center for communication between you and the outside world, allowing you to speak and automatically link to functions that you currently carry on your smartphone. Verbally connecting with transportation systems, power systems, or other buildings will be common. Sentient-appearing buildings will always know your personal preferences as long as they are available on databases accessible to the building, up to and including activity on relevant social media platforms.

The building you are in will always be "on" and aware of who you are and what you are doing. Facial recognition, gait analysis, and other ways of recognizing individual people will be available to all buildings. This scrutiny sounds invasive, but we are nearing that point now. If you have a home device connected to Amazon or Google, it is always listening for its name or a command. When we travel in an urban area, we know that numerous cameras are tracking our movements and our apps on our phones are tracking them too. This trend is accelerating, and we are adding artificial intelligence capabilities to understand and respond to the constant surveillance. This will often appear to be coming from whichever building you happen to be in at a given moment.

When you arrive in a new building, it may greet you by name and instantly know your likes and dislikes, as well as your travel plans and retail preferences. Your friend's house may even check to see if your favorite soft drink is in the fridge. Of course, it will also know if you need to call home. Such awareness can range from great convenience to a massive problem; imagine if a building knew you were wanted by the authorities and responded accordingly! Even sneaking out to buy a birthday gift for a spouse might be difficult to achieve. In any event, sentient-appearing buildings and the networks in which they will be embedded will be a fundamental change in our future urban environment.

Instead of typing commands or manually adjusting settings on appliances as we do now, voice and gestures will communicate with homes and offices, the same as you do with people and pets, and the process will be equally instinctual. Certain buildings may be programmed to have suitable personalities. A hospital may be more "caring" in tone than a business-oriented warehouse. Your home's interface may be changed to fit your whims and may even become aware of your moods and adjust accordingly.

You will be able to delegate elemental decision-making and routine tasks to these buildings, as we do now with such things as programmable thermostats and other sensors that can trigger actions in the mechanical system of the house. If given the authority, home artificial intelligence systems could handle deliveries and visitors, and manage repair and construction work with contractors. It could also communicate with corresponding artificial intelligence systems in other buildings as part of a network of intelligent buildings in the city.

## SHAPE-SHIFTERS

### *Programmable Matter*

In *The Rapture of the Nerds*, a science-fiction novel by Cory Doctorow and Charles Stross, the main character wakes to find himself in a bathroom that wasn't a bathroom the night before. At one point the character notes, "Bits of the bidet are still crawling into position." The kitchen is "...sterile as an operating theater, but that oozes when you glance away, extruding worktops and food processors and fresh cutlery. If you slip, there'll be a chair waiting to catch your buttocks on the way down. There are no separate appliances here, just tons of smart matter."[2]

The most advanced building material of the future may be programmable matter. Generally, this is matter that can change physical properties like shape, color, conductivity, and a few other properties, in response to internal or external programming. Internal programming can be accomplished through the properties of the materials and fluids in the matter, or through such external means as robotic approaches and nanocomputers. Changes can be triggered by the user, or with automatic sensors. The electrochromatic windows described in chapter 2 that become transparent or opaque when voltage is applied are an example of simple programmable matter.

Smart materials, in combination with artificial intelligence systems instructing those materials, will give buildings the ability to alter their configuration on command. If, say, additional rooms are needed in a hotel, or a ballroom is required in place of several conference rooms, the building will be able to change its configuration to meet the needs of the moment. In *The Rapture of the Nerds*, the owner belongs to a House of the Week club, where the house changes itself weekly to reflect a new style.

Of course, all of this has to be done within the limits of the mass available and must necessarily maintain the structural stability of the building. The AI system could also manage contractors, human or robotic, to make large-scale changes in the building. Within those limits, future buildings will be able to make adjustments, conduct repairs, add to or diminish space configurations, and in some cases even change their primary function. They would have the ability to deconstruct themselves at one location and reconstruct themselves in a new location, all controlled by an AI system coordinating with the transportation system, construction robots, and programmable matter.

The ultimate in fluid and changeable material might be utility fog, a term coined by Dr. John Storrs Hall in 1993. Hall's first idea for utility fog was as a nanotechnological replacement for car seat belts. The fog that Hall described was made of smart, communicative, microscopic foglets with multiple arms that could reconfigure themselves into almost anything. Hall saw this utility fog as a replacement for nearly all physical instruments and resources of humankind, including our built infrastructure.

Hall said that, in appearance, utility fog would be more like snow than fog, and that the technology needed to create it is dependent on both the advent of molecular-assembling nanotechnology and the creation of microscopic computers to be built into each foglet. According to Hall, these hardware challenges, though decades from being solved, should be the easy part. The software that controls the fog will be the difficult part, he noted. "The system will have to be capable of keeping track of any changes to the environment and to keep track of you—and this will require incredibly sophisticated simulation, sensing, and interfacing software, and that's going to be tremendously expensive."[3]

Success has been limited. Materials have been invented that can change shapes in response to environmental factors such as moisture. Other efforts focus on origami robots that can shape themselves into tools, walk, swim, and grasp objects.[4]

Animated Work Environment is exploring the idea of an individual room that can reconfigure. They currently have a prototype composed of a programmable robotic environment with a table and eight aluminum panels that can assume six different configurations of screens, whiteboards, lights, and sound, allowing the user to change the structure according to their needs. They hope to develop an advanced version where the tail of the structure can fold around to form a ceiling and floor.[5]

The innovation of smarter materials, along with technology changes such as printed electronics and advanced coatings and sensors, can dramatically change how the interior of buildings look and feel, making them much more interactive.

## Walls

The functions and variability of surfaces such as walls will change. Nanotechnology-level engineering will produce wall materials that have the ability to change shape, acoustical characteristics, color, conductivity, or insulation properties. Printed electronics and embedded information and communication technology will allow wall surfaces of buildings, both interior and exterior, to serve many purposes: screens, sensors, air filters, or power generation.

In addition to sensors being attached to a building, advanced building materials may be able to form sensors when needed, or at least have multisensory coverings. These would be connected to the house AI and/or the internet, according to purpose. This could include coatings such as pressure-sensitive paint to predict the arrival of weather changes. Other sensors may monitor indoor or outdoor air quality.

Researchers at a German university sprayed carbon nanotubes onto a thin substrate film. These carbon nanotube-based sensors are said to detect very small changes in concentrations of gases and could find applications in monitoring the indoor air quality of a building.[7]

Walls, tables, and other surfaces will also become multifunctional through printed electronics. Printed electronics use electrically functional ink that is applied like typical ink onto thin, flexible films. These electronically printed films are already cheap to produce, can be cut to any size, and are robust and durable. With these films, one could create interactive walls to be used for controlling appliances, changing the building environment, or for use as digital, interactive whiteboards. According to Simon Olberding, a developer of such sensors, "By customizing and pasting on our new sensor, you can make every surface interactive no matter if it is the wristband of a watch, a cloth on a trade-fair table, or wallpaper."[7]

Screens will be everywhere. Your walls may serve as TV or computer screens, display Monet's water lilies, pick up where you left off in the book you are reading, or conference in your partner at the office.

The company Meural sells an electronic art screen that hangs on your wall. It can download art and pictures from the internet and from your smartphone. You can cycle through sets of art by waving your hand, or program it to play certain sets at certain times. This type of internet-connected, interactive screen will be built into future walls.

Then there is the power function. Wall paints could not only capture ambient light for electricity but also tap thermal radiation and ultraviolet light for power. The space between exterior and interior walls may not only be insulated, but also filled with material that could store and discharge power, making the wall effectively an out-of-sight battery.

Of course, all these active surface areas will need power. If you have multiple sensors operating, several screens running, and decide you want to change the room from blue to pink, you will need extra power. Energy harvesting, or microharvesting, from on-site vibrations, ambient light, or even thermal gradients could help power the numerous sensors and small processors without having to wire everything to the home electrical system.

Building owners will face choices on how to use exterior walls. You may just want a nice weather-resistant exterior, or you may want to generate power, or it may be desirable to cover the wall with plants to capture carbon, or even use a synthetic carbon capture covering. You may have a commercial building and want part of the wall to be a screen, like Times Square. Sensors networked to the city system may monitor all sorts of urban activity. Again, only one function can occur at a time, so building owners will need to choose their priorities.

## *Self-Repair*

Another way we are embedding information into building technology is through the use of self-repairing or -healing materials. Self-repairing materials provide a significant leap in the life-cycle efficiency of the built infrastructure. Scientists are already working on concrete that can heal itself. Self-healing concrete is being developed using bacteria that trails calcite as it moves into a crack, thus repairing itself. Researchers are also looking at embedding fibers that will shrink concrete in order to squeeze cracks back together, as well as a sort of vascular network that could be used to transfer healing material to cracks where needed.[8] Other materials could detect and repair cracks and flaws, while sensors could alert robotic maintenance to undertake repairs that go beyond the ability of the material to repair or reform itself.

Given the proper authority, a building could contact a contractor for a repair without ever involving the owner. Sensors might detect abnormal fatigue in metals, cracks in concrete, or stress at joints. The building AI would deduce a repair is necessary, solicit bids, approve proposals, check the work, and arrange for payments. If given a budget, work parameters, and financial authority to execute work on the owner's behalf, the building could handle its repair as needed.

## *New Materials*

Buildings of the future will utilize composite materials that combine two or more materials with significantly different physical or chemical properties. Of particular interest are polymer matrix composites that can combine reinforcement materials such as carbon, glass, or natural fibers with particulate material such as sand, talc, or recycled glass. These composites can make building materials that are extremely durable, lightweight, and energy-efficient, and very flexible in design because they can be cast, molded, sprayed, and shaped through various processes.[9]

A certainty for the far future of buildings is that they'll be fashioned from thinner, stronger materials than those presently used, continuing a multicentury trend as builders moved from stone to steel I-beams to more advanced materials. Carbon, in various forms, can be a valuable building material. Much of our built infrastructure of the future could be built from carbon that has been mined from the atmosphere.

Builders can replace steel, copper, and aluminum in many functions using carbon nanotubes.[10] Carbon fibers are common today in bicycles, automobiles, and airplanes and will be a widespread building material in the future, in many different forms.

Another form of carbon is graphene. Already used today in such things as high-end tennis rackets, graphene is a pure form of carbon only one atom thick, 100 times stronger than steel, and as rigid as diamond. Such properties obviously make graphene a good candidate for future building materials. Graphene is also highly conductive and can be made to be flexible, so it might be used to build both strong walls and malleable panels that could change color and morph into television displays.[11] Though quite expensive today, graphene's expanded production and demand could cut its cost by 70 to 80 percent. While graphene holds great promise, hurdles related to large-scale manufacturing of the material remain.

As amazing as graphene is, another carbon-based material, carbyne, might be even thinner, lighter, and stronger. Although it has only been observed and not yet synthesized, carbyne is possibly the highest energy state achievable for stable carbon, a one-dimensional material that, according to its inventors, should have the highest tensile strength of any known material and a tensile stiffness nearly three times that of diamond.[12]

The sustainable materials technology that we saw in the previous chapter will also be present, such as smart windows and phase-changing insulation. Particular emphasis will be placed on technologies that can respond to changes in the environment.

## *Upcycling*

The good news is that the ease with which buildings will be constructed, reconfigured, and torn down will enable greater flexibility, which means the buildings will find greater utility. The bad news is that doing so reinforces the popular mind-set of a disposable culture. That is, we live in a society that already has a widespread problem with using goods once before disposing of them, thus creating vast volumes of waste. Adding buildings to the list of disposable goods could be devastating. However, adding buildings to the list of recyclable goods could be transformative in positive ways because they become a natural destination for the reuse of our resources. Along the way, the type, functions, and locations of buildings may change significantly. The solution to this concern would be upcycling.

Recycling can be transformed into upcycling, a term coined by William McDonough and Martin Braungart in their book *The Upcycle*. They discuss designing and engineering technology components to be more interchangeable as "technology nutrients" and consistently reform as an upcycle, where the combination is more efficient and sustainable than the last one.[13]

Buildings that are constructed in modules and new materials, and designed with upcycling in mind, can be continually reused in new forms. There will be ways to use both the organic and inorganic material waste generated in a building as feedstock for 3D printers, composters, or other means of transforming waste.

Buildings of the future will take zero-waste to a new level. Reconstructing the recycled material with advanced construction methods could lead to very fast and efficient housing construction. And as sustainability and upcycling principles take hold in the manufacturing sectors, some companies might even stop making products that cannot be recycled, as IKEA has decided.[14]

## AUTOMATION

Today we are used to a certain degree of automation around us at home and at the office. The thermostat changes the temperature, the doors open automatically in the office building, and security lights go on at night. In future buildings, much more will be going on around you, some of it overt, some of it behind the scenes. You may hear the clothes dryer or dishwasher running, and you did not start them. Windows and floors may be cleaned on a schedule.

The future buildings will use a combination of built-in equipment and mobile robots to maintain the building and serve the occupants. We already have Roomba that works autonomously to vacuum, and pool cleaners that move about on their own, but the capabilities will become more sophisticated. Some building-integrated automation may handle the laundry room and kitchen functions. In the future, we may just leave things to be cleaned, like dirty clothes or dishes, in the appropriate place for the automation to handle. (Author's note: Some of us do that today regardless, but nothing happens.) This can apply to other daily chores as well, such as leaving our electronics that need to be charged on the appropriate surfaces for wireless charging.

The kitchen may handle food preparation robotically. There are currently commercial robot restaurants where you can watch your food being prepared by robotic arms and bins and burners while you wait.[15] Such robotic food preparation could be incorporated into a modern home kitchen, although we should probably not expect haute cuisine at the beginning.

Laundry folding devices have been under development for years. Two robotic devices are currently available, but they are still rather large and expensive. Relatively inexpensive and effective devices will evolve in the future. This is one area where robotic arms or chutes may be built into the structure. The laundry room becomes not just a location for the washer and dryer, but also a full laundry service.

And the fully automated house may follow you around inside. Currently, in some buildings the room lights go on and off as you move around due to motion sensors. Now imagine a more fully aware building environment that follows you around and anticipates what you need based on your past behavior. Not only will appropriate lighting adapt to your location, but also perhaps a place to sit will flow out of the wall and a nearby surface will be made available as a screen. Home services will follow you around as the building learns your behavior. In fact, your home may often anticipate what you are going to do next even before you do.

## AUTONOMY

The really interesting question is, will these buildings become fully autonomous, with control over their physical structures? There are certainly a few indications that they may.

First, the building will be able to meet its power needs through a combination of on-site power generation and purchase of power from the grid. Building-integrated photovoltaics, and several other small devices, can very effectively convert the natural flows of energy in and around the building to power. The AI can work with the grid to purchase clean energy and store or sell power as needed, predict power loads for the grid, and manage all the power loads for the building.

Second, the building will either be mobile itself, or the AI can arrange for relocation. The trend toward trailers or other forms of mobility will increase. We have seen a proliferation of the types of small buildings that are now trailers. In addition to RVs, there are now mobile offices, clinics, libraries, educational displays, food trailers, and much more. But there may be a new way for a building to relocate in the future, thanks to trends in modular construction and the use of robots in construction.

In the future it will be easier to dismantle a building and reconstruct it at a different location. This is especially the case if the building was designed for reassembly elsewhere and modular construction was used. Offices, emergency shelters, and clinics could be dismantled and reconstructed quickly where needed—say, in a disaster zone. Homes might be able to move with the owner to a new lot in a different city.

When all these properties are considered together, the end product should appear autonomous and self-sufficient.

### Smart Buildings and Dumb Buildings

Because buildings last for decades, turnover of capital stock takes a long time, and many existing structures may be kept for practical or historical reasons. The future built infrastructure, therefore, will be a mixture of the old and the new, just as today. This means that there will be both "smart" buildings and "dumb" buildings. Older buildings may be retrofitted to enable basic communications and sensors, but remain the same otherwise, unable to change their composition or structure.

These older buildings may have a façade from an older structure, but inside they will have the latest wireless communication and an artificial intelligence system installed in a nonobtrusive manner. Even so, the building would still not have the "smart material" and advanced automation and robotics of a modern building. The disparity between "smart" and "dumb" buildings will grow ever greater over time.

## NEW ROLES FOR BUILDINGS

### Buildings as Health Monitors

What if your home could read your temperature and heart rate? Sometime not too far in the future, it will likely do this and a lot more. It's anticipated that all manner of buildings will act as health monitors. Motion sensors, for instance, will monitor your location and alert health-care responders if you've fallen. As our population ages, many families are retrofitting their homes with motion detectors, sensors, alarms, and other devices to allow senior citizens to remain in their homes as long as possible. These smart homes can also arrange for transportation to visit friends, doctors, or stores, thus reducing isolation for the elderly and improving activity and therefore health.

Smart homes might be built with the most sophisticated health monitoring equipment, perhaps replete with a laboratory toilet. The lab toilet could analyze waste for diabetes, kidney disease, nutritional deficiencies, and many other issues. Sensors in our refrigerators would alert us when foods spoil or chilled medicines expire. While such devices, along with health monitoring homes, may sound strange or intrusive today, they'll more than likely be a part of life in the near future.[16]

In many ways, these on-site health sensors will be a form of distributed medicine. Just as we are moving away from far-flung, large-scale power plants to on-site power production, from large-scale water treatment plants to on-site water harvesting, and from large-scale factories to on-site 3D printing, we may also reduce our use of large-scale health-care systems (e.g., hospitals) for on-site diagnosis and treatment. Hospitals should be reserved for specialized procedures that cannot be done at an outpatient facility.

Hospitals themselves could become some of the most automated buildings. Today, a stay in the hospital may involve robot-assisted surgery in an operating room sanitized by robots. Your medicine and meals may be delivered by robots. Your blood labs are automated, and patients may even receive a fuzzy animal-like robot to comfort them afterward.

In the future, robotics and automation will play an even bigger role in medical diagnostics, surgery, lab testing, nursing care, admissions, insurance, and payments. Hopefully there will still be human doctors and nurses to guide us through a hospital stay, but much of the care will be robotic and automated.[17]

## Urban Farms

Cities of the future are often imagined in conjunction with a proliferation of urban farms, either as individual buildings or space set aside in a multipurpose building. Many modernistic visions include indoor agriculture as part of the self-sustaining nature of multiuse skyscrapers. Others imagine single, multistory, vertical urban farms producing both animal and vegetable foods. Mixed-use designers often call for these vertical farms to have lower-level retail space where the food produced above can be sold.

Growing our food in controlled settings might improve efficiency and help prevent contamination of food and the environment. Vertical urban farms would also, according to proponents, help negate the need for "food miles," or the costs and pollution related to shipping foods. Food would also be fresher since it will take less time to move it to the consumer.[18]

Others, however, see vertical food farms as a rather pie-in-the-sky idea. Horticulturist Cary Mitchell of Purdue University says, "The idea of taking a skyscraper and turning it into a vertical farming complex is absolutely ridiculous from an energy perspective."[19] Instead, Mitchell thinks the future of urban farming will be in the suburbs, in huge, flat warehouses where land and power are cheaper.

The flatter design of these warehouses negates the need to add as much artificial light compared to a multistory layout, saving considerable energy, money, and stress on the environment. Instead of the traditional fluorescent lighting used in most of today's greenhouses, urban farms of the future will likely use a mix of red and blue LED lights because those are the wavelengths of light that plants need most.

Not only are LEDs more efficient in general, but also they can be tuned to give off a small range of necessary light, and it has been shown that this method works as well for growing plants, if not better, than using the entire light spectrum. "Twenty years ago, research showed that you could grow lettuce in just red light," Mitchell says. "If you add a little bit of blue, it grows better."[20] An additional benefit is that LEDs operating for a small part of the spectrum are cooler. That means they can be closer to the plants, which means they can operate with lower losses and less energy consumption.[21]

As LEDs get more efficient, future farmers could be growing vegetables in completely enclosed buildings, not having to worry about temperature or pests. For example, iBio Biotherapeutics has an indoor plant factory in Texas that grows 2.2 million plants, stacked 50 feet high, under blue and red LEDs.[22] Plenty, a firm that specializes in vertical farming, has announced plans for a 100,000-square-foot facility near Seattle, and Chicago is home to The Plant, a former pork processing plant that is now home to several indoor farms.[23]

However, with the exception of marijuana-growing operations, current urban farms are having difficulty competing with traditional vegetable farming because of electricity and labor costs. It is hard for a high-tech growing operation to compete with vegetables grown in dirt with free sunlight. There have been several indoor farming failures such as FarmedHere and PodPonics. Google gave up on indoor farming because it could not grow staple food products like grain and rice economically. Indoor farmers need electricity costs to be around 3–4 cents per kWh, while the national average is around 10 cents.[24]

Even so, if climate change, loss of farmland, and growing population stress the food supply, then urban farms may become a vital part of our food supply. There also may be a market for food grown under very controlled conditions. And many buildings, both home and office, may feature grow rooms or even grow floors.

### Urban Forests

Although commercially growing food in urban towers is proving difficult, another type of urban greenery seems to be taking off. Using terraces to grow trees, shrubs, and plants is a growing (pun intended) architectural movement.

The first vertical urban forest grew on two residential towers in Milan, Italy that hosted 800 trees, 4,500 shrubs and 15,000 plants—the equivalent of 20,000 square meters of forest.[25] A few dozen buildings around the world are planting forests on high-rise buildings, making a substantial addition to the urban forest, with the improved air quality and environmental benefits associated with trees.

In France, a multiuse, 54-meter tower will be covered with 2,000 trees and designed to allow natural illumination and ventilation to the apartments.[26] This is not just a luxury for the rich; the Trudo Vertical Forest will contain 125 social housing units. Each apartment will have the exclusive benefit of 1 tree, 20 shrubs, and more than 4 square meters of terrace.[27]

Several urban forest towers are being planned and built in China. The Nanjing vertical forest has two multipurpose towers that will hold 1,100 trees and 2,500 plants and shrubs. Several other Chinese cities are planning similar towers.[28]

Then there are the 18 Singapore Supertrees. Each Supertree is a reinforced concrete trunk wrapped with a steel frame and a canopy shaped like an inverted umbrella. Planting panels on the sides provide for a living skin, and the 18 Supertrees host more than 162,000 plants from more than 200 species. Some have PV cells that light up the Supertree at night.[29]

Architects are now seeing terraces and rooftops as vital, outside workspaces for employees. A planted terrace or roof space with electrical outlets and Wi-Fi can be a very attractive amenity. Technology giants are all following the trend: Facebook has a nine-acre park on top of its building in Menlo Park, California; Microsoft has tree houses in Redmond, Washington; and Samsung has some terraces in its San Jose, California office devoted to quiet contemplation, while another is a putting green.[30]

One of the latest trends involves covering urban freeways with deck parks. Dallas, Texas covered part of the Woodall Rodgers Freeway with a five-acre park and broke ground in 2018 on another deck park. Many other cities are now looking at doing the same, with Atlanta considering three such projects.[31] With outdoor terraces, roof gardens and deck parks being added, cities could start to look more like tree-covered mountains than cold, steel urban jungles.

## Specialized Facilities

The combination of new materials, artificial intelligence, and advanced automation will increase the trend toward highly specialized buildings. We have seen this trend for years in everything from sterile clean rooms to specialized data center buildings. This will expand further for housing electronics and specialized manufacturing facilities designed around particular products.

Data centers supporting the internet are major energy consumers. They are designed to handle the heat from the computers and reduce electrical loss, with multiple layers of redundancy in the power source. These buildings will tend to be located in cooler locales with cheap energy sources. Large companies may want to colocate the data centers with renewable power sources as much as possible. Through the use of exotic shell materials, advanced insulation and heat recovery, these electronic shelters will ensure the proper temperature and humidity for optimal operation.

Where will manufacturing take place in the future? Some think the proliferation of 3D printers will distribute manufacturing throughout homes and offices, much the same way that solar panels are starting to distribute power generation throughout the built environment. Homes and offices will manufacture some of their own needs, and existing manufacturing facilities are already becoming some of the biggest users of 3D printers as they print parts on-site. We may even see a proliferation of small backyard manufacturing sheds designed to hold printers and materials. Someday homes and offices will take waste material produced on-site and use it as 3D printer feedstock, producing useful items on the spot.

3D printing is already showing up in local office supply stores that offer printing services. This mode of additive manufacturing will transform manufacturing, but there will still be a need for large-scale manufacturing facilities for making the underlying construction modules in high volume, and manufacturing large volumes of a single product will still be faster and cheaper than the slower 3D printing.

These facilities might use smart materials, internet-connected manufacturing tools, and other new elements of smart manufacturing to become highly specialized buildings in and of themselves. These buildings would have environmental controls, distribution facilities, and tools built into the structure, specifically designed for the product being manufactured. There would need to be highly controlled and specialized facilities for some nanotechnology manufacturing, where clean rooms and containment are issues.

Fully automated warehouses might become commonplace in our future. Already we see fulfillment centers from Amazon that are essentially machine-filled spaces operating effectively and nearly autonomously, with a few humans at the center. As we will see in coming chapters, the entire freight delivery system may become automated, with automated warehouses handling the transition between national and local delivery systems.

## Buildings as Power Plants

In the near future, most new buildings will be generating at least some of their own power. Major league sports see the solar value of their large stadiums. As of 2019, about a third of all the stadiums in the NFL, MLB, and NBA leagues are fitted with solar panels.[32]

Lincoln Financial Field, home of the Philadelphia Eagles, hosts PV panels that generate 3,000 kilowatts, or 3 megawatts.[33] That is enough electricity generated in a day to power more than 2,000 homes for a year.[34] We think college and high school stadiums around the country will follow this trend as solar prices drop.

Stadiums around the world are following the same trend. The 2014 World Cup kicked off in Brazil in the first World Cup stadium powered by solar energy.[35] Japanese architect Toyo Ito designed the recently completed 55,000-seat arena in Taiwan, powered by 8,844 solar panels. The stadium can generate 1.14 gigawatt hours annually, which is enough to power 80 percent of the surrounding neighborhood.[36]

Many other types of buildings today could essentially become power plants. Warehouses, reservoirs, strip malls, and schools—any building with low energy loads and a lot of sun-facing surface—could operate as power plants most of the time. Japan built one of the world's largest floating solar plants in a reservoir, which helps reduce evaporative losses of water while also generating electricity.[37] In 2017, China surpassed Japan and built a floating solar farm with enough capacity to power 15,000 homes.[38]

Schools are especially good targets for solar generation. Their large, flat roofs are perfect for PVs. Likewise, school parking lots could be covered with PV canopies and become power productive. Even better, schools are out for most of the summer in the United States, so when the panels are achieving peak production, the electricity can be sold to the grid, generating revenues for the school.[39]

Large surface parking lots, such as at malls, airports, and big commercial industrial sites, also have great solar energy potential. When the right support structures are cheaply manufactured, it will be easier and cheaper to cover surface parking than to deal with the varied commercial rooftops.

More structures will be built with building-integrated photovoltaic (BIPV) materials. Roofing shingles, power paints, power windows, and sheets of solar panels that can be spread over almost any surface will greatly expand the amount of energy typical buildings generate from the sun.

Solar roofing tiles have been in development for years. CertainTeed, a traditional building materials manufacturer, sells thin solar shingles that can replace or simply lay atop existing asphalt roof shingles or tiles. The manufacturers claim these systems can save homeowners 40 to 70 percent on their electric bills.[40]

Tesla is now producing a solar roof tile that is very popular in regions with high electric rates. Tesla started selling the tiles in May 2017, and within a few weeks orders were backlogged well into 2018.[41] They began installing the first Tesla roofs in the spring of 2018 in California.[42]

The near future should also see more buildings harnessing wind power and integrating microwind-power production into the building design. Buildings of all kinds will take advantage of on-site thermal and kinetic energy flows.

And as technology improves, our buildings will harvest more of their own solid waste. Thus, in the future, we won't just see more net zero energy buildings (ZEBs), but also ZWBs: zero waste buildings.

Buildings themselves may be used as large-scale energy storage units. New materials, especially when engineered to nanotechnology specifications, could greatly expand the possibility of structural materials that could also store energy, in addition to batteries.

## Buildings as Carbon Filters

Earlier in the chapter, we introduced the idea of buildings harvesting carbon from the air. Though the idea of homes and offices filtering carbon from the air and using it as building material may seem far-fetched, it isn't. Switching buildings from being a source of carbon to a sink for carbon would be a major transformation. In fact, we don't have to look far to see inspiration for such systems. For example, ivy growing on walls and buildings, while decorative, actively takes carbon out of the air. There are rocks in Oman that take carbon out of the air, so if those types of rocks are converted into building materials, then our homes will provide that service.[43]

Buildings may become valuable assets for the space needed to mine atmospheric carbon. Covering outside building surfaces with either inorganic filters, plants, or other organic material to breathe in carbon dioxide could be an effective way to capture and store carbon in the future. In 2013, a house was built in Germany with a bioadaptive façade designed with algae that grows faster in bright sunlight, offering cooling shade while at the same time producing biomass that can subsequently be harvested and used to power the building.[44]

Covering building exteriors with plants is a good way to combat climate change, but synthetic carbon capture may be necessary in some climates, and artificial leaves on artificial trees can capture much more carbon than natural trees.[45] We may see a mixture of natural and artificial landscapes in the future. Buildings and cities may still be green, but it won't all be natural.

Other façades work much like animal skins, adjusting shading and venting in response to ambient humidity, temperature, and light—in a sense making it like a breathing skin. The capture of greenhouse gases may even extend to the furniture in the building. *Popular Science* gave its 2014 Innovation of the Year Award to Newlight Technologies for their development of AirCarbon, a new plastic made using captured methane to form a polymer that performs the same as oil-based plastics. Companies are already using this new plastic for such items as desk chairs.[46]

## *Multipurpose Skyscrapers*

Many envision a future filled with skyscrapers. Lots of them. If we transported ourselves back a century ago and asked a futurist at the time where skyscrapers would be built, they surely would have said North America, Japan or Europe. Today, when you ask futurists the same question they say that the world's next great skyscrapers will be built in China or the Middle East. In fact, some believe that China will build 20,000 to 50,000 new skyscrapers over the next few decades, located in more than 220 cities with populations exceeding a million people.[47]

But these won't be just tall buildings like the ones we have today. Many believe the skyscrapers of the future will be multifunctional minicities, not merely housing residents but also creating energy, offering on-site food production, commercial and park space, and much more. In dense city cores, building up has been the norm for more than a century. In the sprawling cities of the U.S. at the turn of last century, a "New Urbanism" sparked a resurgence of mixed-use condos and apartments. While most modern buildings are still single- or limited-purpose, many see the minicity tower as the predominant building mode of the future. Shawn Gehle, a principal at Gensler, a global design and architecture firm, says, "The Shanghai Tower we're currently completing in China is effectively a city within a building."[48]

Imagine a building connected to a smart power grid, but one with solar PV paint applied where possible, phase-change materials functioning as heat-recovery agents, an integrated micro-wind-production system, a water collection and recycling system, a building membrane or on-site filtration system that converts carbon dioxide into oxygen, heat recovery windows, an algae façade or moat to produce biofuel and clean water, a recycling center, and urban vegetable farms. Connecting the building to the rest of the city are cable cars, pedestrian bridges, an underground transport hub, and a car- and bike-sharing center. The whole thing would be built of modular building components that could be retrofitted, moved, and replaced, all by an on-site robotic assembly and maintenance crew.[49]

Some imagine these tower cities as not just multiuse, but flexible and responsive even to the specific needs of individual dwellers, "essentially functioning as a living organism in its own right—reacting to the local environment and engaging with the users within."[50]

Buildings with solar and wind production, urban farms, recycling centers, cable cars, and solar PV paint are probable in the not-so-distant future. But the authors see an issue with claims of the self-sufficiency of such structures. For the same reasons it's presently impossible to build tall, skinny, high-energy-load zero energy buildings, most urban skyscrapers won't likely be self-sufficient, and will require power from the electric grid.

The attachment of multipurpose sensors in and on buildings, especially in an urban environment, can give a city a continuous monitoring of sights, sounds, pollution, and much more data. Several academic institutions are now researching how the collection of such data could affect pollution, traffic congestion, crime, property values, and much more—including the privacy issues resulting from such pervasive surveillance.[51]

### Climate Change Response

The buildings of the far future will make several changes to accommodate climate change. The locations will change in response to rising seas and increased heat. Droughts, floods, wildfires, hurricanes, and tornados will change the regional built infrastructure accordingly. Some buildings may excel in rainwater capture in arid regions, while coastal homes may be designed to float as the water rises.

There may be a movement toward relocating more building underground. If you travel west from Taos, New Mexico, across the Rio Grande River, you'll soon encounter a colony of odd-looking homes built into the ground, usually topped with solar panels. The hippies who first built these Earthships did so for the same reason early man built homes embedded in the ground: because they stay warm in winter and cool in the summer.

Earthships are also readily powered by solar PV, and are desirable because they're more integrated within the greater ecosystem than today's traditional housing. What's more, by definition, building material is plentiful, cheap, and easily accessible. Future man may construct many buildings in earth-sheltered homes as well, both for efficiency, and in response to expected climate change. Some may even choose to build, live, and work fully underground for the same reasons.

Construction will be hardened in some areas to withstand harsh weather conditions. Other changes will include more fire-resistant houses, advanced rainwater collection, local electrical microgrids, and advanced drainage design. In general, although buildings may become superintelligent and filled with advanced technology, much of our future building modifications will be in response to an increasingly harsh environment.

## CHAPTER 4 SUMMARY

1. Advanced computation and communication technology will give buildings the ability to interact with us and control their structure as if they were people.

2. Building operations and maintenance will become much more automated and some buildings may reach a degree of autonomy, able to power, repair, and even relocate their structure.

3. Buildings will take on new roles in our lives, becoming a monitor of our health, a principal mode of communication, and a personal assistant.

# PART 3

# THE FUTURE OF
# TRANSPORTATION

Mary was running late. Yet again. Her phone reminder of when her plane was leaving was like some kind of electronic rebuke. The phone app said the car was en route, but she wasn't ready yet. Robots suck, she thought, as she wheeled her suitcase to the curb to await the electric self-driving ride to the airport.

In the airport, she grabbed a cup of coffee and a donut on the run. The smell of the oil at the donut kiosk reminded her of the vaguely organic "French fry" aroma that permeated airplanes nowadays—the byproduct of the biofuel that powered most commercial aviation.

The flight to Boston was on a sunny and clear day, and as she looked down at the platoons of driverless trucks on the highway, she wondered what freight they were carrying. As the plane neared landing, she wondered the same thing about the nuclear-powered container ships laboring in and out of the Inner Harbor. Incongruously, working alongside the automated vessels were commercial sailing ships. Cargo that wasn't time sensitive was often sent to market under sail. After two centuries, Boston Harbor once again was teeming with sails.

After leaving the airport, she noticed that the electric scooter plague had gotten even worse. This is how mankind ends, she thought wryly—not by a nuclear war, but by being run over by a 15-year-old on a scooter. Or worse, by a 70-year-old in a suit on a scooter.

She unpacked and toppled into bed, leaving a wake-up call with the room's computer. She had time for a couple of hours of shut-eye before she met her first client for drinks.

# ENERGIZING TRANSPORTATION

**W**hen the petroleum-fueled internal combustion engine appeared on the scene, it was the solution to the environmental problem of the time: horse manure.

At the end of the 19th century, horse manure in cities had become a major health hazard, and the smell had become overwhelming. Dead horses sometimes lay for days in the streets, and horse flu could practically shut down the economy of the east coast. City planners said that urban populations had reached their maximum because cities could not manage more horses.

Oil solved that problem. Some of the first cars were electric, but batteries at the time could not compete with the range, power, and cost of an internal combustion engine fueled by gasoline. In just a few decades, horses almost disappeared from the urban landscape. The health hazard and the smell were gone, and the number of cars had not yet approached the point where air pollution had become an issue.[1]

Petroleum dominates our transportation system, and the solution to an earlier environmental problem is now the major contributor to the number one environmental crisis of our time: global warming. As Kurt Vonnegut wrote, "Dear Future Generations: Please accept our apologies. We were roaring drunk on petroleum."[2]

We are finally moving toward other options for powering transportation. Major players in the fossil fuel industry assume that peak oil demand will occur sometime in the next several decades. British Petroleum, Royal Dutch Shell, Statoil, Total, and the U.S. government all see oil demand peaking between 2025 and the 2040s.[3]

This is a reversal from the conventional wisdom of the late 20th century that peak oil supply would be our problem. Rather, we have found that there is an abundance of oil, gas, and petroleum resources, and we are now anticipating peak oil demand instead.

This chapter is about the many ways we are kicking the oil habit. There are at least three major areas of change underway in our transportation sector. First, there is the cultural change in the way we own or use vehicles daily. Second, there are fundamental shifts in transportation technology. And finally, alternative fuels are capturing the fuel market.

But there are obstacles that will make this transition slower and more expensive than we want. An abundant oil supply, corporate lobbying power, infrastructure needs, and the rate of vehicle turnover all mean that this transition is going to take more than a few years. And some transportation sectors will be difficult to change. In particular, freight movement, large commercial aviation, and international shipping will be difficult and expensive to wean from petroleum.

Nevertheless, we are seeing a remarkable change in our use of petroleum to move our stuff and ourselves around the world. And the most dramatic change is to the culture of the car.

## PEAK CAR

We two native Texans recall vividly when a driver's license and access to our parents' cars were irresistible symbols of liberation and freedom. The open road, the keys to the highway—these were the subjects of songs by Chuck Berry and Bruce Springsteen (best heard over the car radio, of course)—TV shows like *Route 66,* and movies such as *Easy Rider* and *Thelma and Louise* relied heavily on the allure of a set of wheels and the open road. Michael's mother took Route 66 from St. Louis to Los Angeles for the opening of Disneyland in 1955.[4] And there weren't a lot of ways to get around the sprawling Lone Star State that didn't involve an automobile.

But now the demand for personal automobiles may be reaching a saturation point, and automobile ownership may be falling out of favor. Ride-sharing apps like Uber and Lyft have made car ownership less of a necessity in the cities. And car-sharing services like Zipcar have taken root among consumers enticed by the idea of a "disposable" automobile. Since these business models charge on a per-trip basis, they can also reduce the number of short trips or other travel not seen as entirely necessary.

In addition, driving is showing less appeal to the generation now coming of age. According to *Bloomberg Businessweek,* only 26 percent of U.S. 16-year-olds earned a driver's license in 2017, and "the annual number of 17-year-olds taking driving tests in the UK has fallen 28 percent in the past decade."[5]

Urban planners are also working to pry individuals out of their cars. Municipal transit services, urban hike-and-bike trails, designated bicycle lanes, "smart city" planning, and transit-oriented development are all reducing the number of people traveling to work and play in their own vehicles. The recent national lockdown has spurred many large businesses to set up their employees to work from home. They have found that it works fairly well, and many will not return to packed downtown offices.

Also, the "first mile/last mile" of travel in a car is being substantially replaced with two-wheelers. In addition to the bicycles and motorcycles we have become accustomed to, we are now seeing a flood of "micromobility" vehicles. People are weaving among us on e-bikes, electric scooters, and electric skateboards.

Finally, regulatory actions around the world are restricting vehicles allowed in big cities. Manhattan recently joined London, Stockholm, and Singapore in putting a price on driving through their densest areas. This policy is known as congestion pricing, and studies have shown that it can significantly improve air quality as a handy side benefit to reducing congestion.[6]

For all the reasons above, we may be witnessing "peak car" in some developed countries. According to a March 2019 article in *Bloomberg Businessweek,* "Auto sales in the U.S., after four record or near-record years, are declining this year, and analysts say they may never again reach those heights." That scenario is not limited to the US.

Bloomberg notes that IHS Markit, a research firm, sees the huge impact of mobility services in China. "Auto sales there plunged 18 percent in January (2019)…as commuters rapidly embraced ride-hailing."[7]

## TECHNOLOGY CHANGES

### *Electrification*

Electric cars made their debut in the nineties…the 1890s. James Prescott Joule, a famous physicist for whom a unit of energy is named, first anticipated them in 1839. Since then, dozens of models and designs have come and gone. But the resurgence of the electric vehicle (EV) is strong today and electric cars seem destined to dominate our local transportation.

Electric vehicle sales have steadily increased in the U.S. and are soaring in China. The International Energy Agency predicts there will be 130 million electric light-duty vehicles on the road by 2030.[8]

A global coalition of countries has the aspirational goal of electric vehicles taking 30 percent of the market share by 2030.[9] Others have projected much more ambitious targets for the transition. Incentives by the auto industry and government (via rebates and tax breaks) will hasten the increased adoption of electric vehicles.

Cars, sedans, vans, and most trucks will be electrified in the coming decades. These so-called light-duty vehicles (LDV) account for 59 percent of the oil used for transportation in the U.S.[10] Whereas previous attempts to wean Americans off gasoline and diesel fuel have failed, the takeover of this market by electricity seems inevitable.

At the core of this transition is the relative efficiency of electric motors compared with internal combustion engines. According to the Department of Energy, electric vehicles convert 59 to 62 percent of electric energy from the grid to power the wheels. Conventional gasoline engines, by comparison, only convert 17 to 21 percent of the energy stored in gasoline to propel a vehicle.[11] Furthermore, since the motors are directly connected to the wheels, there is no wasted consumption when stopped or coasting.[12]

The main obstacles so far to adoption of electric vehicles have been the up-front cost of the car, and anxiety over the range and charging locations. These problems are quickly being solved. It's estimated that by 2025, an electric vehicle and a gas-powered car will cost roughly the same.[13]

Once the price of an EV is the same as a conventional car, the consumer movement to electric should take place quickly. Electric vehicles have many advantages over conventional cars, including fuel cost, operational cost, a better driving experience, and a lower environmental footprint.

Electricity is cheaper than gasoline. As of March 2020, the average price of a gallon of gas in the U.S. was $2.25, according to the Department of Energy, and the average cost of an electric "e-gallon" was around $1.15. An e-gallon is the amount of electricity it takes to power an electric vehicle for the same distance a similar vehicle would go on one gallon of gasoline, and its price reflects the average electricity price, varying by state.[14]

In addition to being cheaper to fuel, the EV is simpler and much cheaper to maintain. While your standard internal combustion engine vehicle contains about 2,000 moving parts, electric vehicles have as few as 20.[15] There simply is less to maintain and fix.

Electricity also has a fueling infrastructure advantage over other alternatives to petroleum. Power plants typically have excess capacity in the evening, and the distribution system already exists. There are currently tens of thousands of charging ports and many more underway, compared to fewer than a thousand natural gas stations, and fewer than a hundred hydrogen stations.[16] Biofuels can use existing gas stations, but must build new refineries to expand their market.

And the final reason that EVs will take over this market is that electric cars just provide a better driving experience. They are absolutely silent in operation, and the vibration is gone as well. The acceleration from 0 to 60 is quicker than a regular car, since the electric motor provides full torque from a standstill.

When we move from petroleum-driven engines to electric motors, we open the transportation sector to a wide variety of clean, renewable sources of energy. We can bring the power of solar cells, wind turbines, dams, and nuclear reactors to move our vehicles. That diversification of transportation fuels is good news, partly because the renewable energy transition taking place in the power sector can also be used to clean up our transportation emissions.

While the emissions from manufacturing electric vehicles are higher than from manufacturing similar-sized internal combustion engine vehicles, the overall reduction in emissions over the lifetime make up for it several times over. The emissions from driving an electric vehicle depend on the fuel mix that's powering the grid it's plugged into, but on average in the U.S., battery electric vehicles produce less than half the global warming emissions of a regular gas-powered car.[17]

Furthermore, the emissions from an electric vehicle decrease with age because the power sector is getting cleaner year after year as we continue to close coal plants and open new renewable energy generation. This is in contrast with conventional diesel and gasoline engines, whose performance degrades with time, yielding more emissions year over year.

Finally, although some emissions from an electric vehicle may just be moved from the car to the power plant, it is much easier to control a relatively small number of smokestacks compared to millions of tailpipes.

But electric vehicles do have environmental disadvantages. In their manufacture, a significant amount of greenhouse gases are emitted. Mining rare earth metals can be a toxic and environmentally damaging process because of the chemicals used to extract the ore and the radioactive elements that are dislodged from the earth in the process.[18] A study from the firm Arthur D. Little concluded that battery electric vehicles have three times the human toxicity potential as gasoline vehicles, and that toxicity could grow to five times by 2025 as the battery packs increase in size.[19]

Lastly, there are substantial environmental impacts of large-scale battery production and disposal. Reuse and recycling of the battery packs is an important part of moving toward sustainable use of electric cars. There has been discussion about using older car batteries as a storage source for electric utilities, and as the first major round of electric vehicle batteries are wearing out, recycling programs are starting to emerge.[20]

The EV battery still has a lot of power left after it can no longer be used in an electric vehicle. The batteries can be used in several different ways. Their greatest use might be as energy storage for solar panels. Toyota is already teaming with 7-11 stores in Japan to reuse the car batteries to store rooftop solar power for some store operations. In England, old EV batteries are being used in homes and schools. In an apartment complex in Sweden, reused Volvo batteries, charged by solar panels on the roof, power the elevators and lights in the common area.[21]

But the reuse programs just delay the inevitable need to recycle the components of the batteries. It is still far cheaper to mine lithium and other rare earth metals than to recycle batteries. Fewer than 5 percent of the lithium batteries are being recycled.[22] So even though there are many processes to handle the recycling, the economics are not yet favorable. This problem needs to be addressed, and we expect that eventually there will be multibillion dollar markets in battery reuse and recycling.

The heavy-duty trucks and semis that haul our goods and food across the country are also going to be electrified. Moreover, electric buses are expected to take over half of the world's bus fleet by 2025, from 386,000 in 2018 to 1.2 million by 2025.[23]

Truck makers have several efforts underway to make tractor-trailers electric, and we should see electric long-haul trucks on the highways in a few years. But there still is the basic problem of the size and weight of the battery. It can become the bulk of the load, at the expense of freight. On the other hand, the infrastructure for highway charging stations for heavy-duty trucks could be set up relatively quickly.

Another option to electrify heavy-duty vehicles is to use fuel cells. Hydrogen-powered fuel cells are already used in niche trans-portation vehicles, such as forklifts. Some truck manufacturers are also testing large fuel cell trucks for use at ports and other locations.

We are also seeing the electrification of two-wheeler vehicles. This is a big deal in transitioning the developing world off petroleum. Cities in many parts of the world are not dominated by cars, but are thoroughly congested with motorbikes and motor scooters—all gasoline powered. The transformation of this sector of transportation will help solve our global warming problem, and immediately start to clear the air in some of the world's biggest cities.

China has been leading the world in transforming two-wheelers to electric. China now has 250 million electric two-wheelers, and most motorcycles sold are electric. China has placed restrictions on gasoline motorcycles, and the electric versions qualify as bicycles and can use bike lanes and do not have to register.[24] It is unclear if electric two wheelers will be as successful outside China without such regulations.

Many countries have proposed banning internal combustion engines for transportation, further reducing oil consumption. The United Kingdom announced an ambitious plan to put its entire transportation market on an all-electric status by 2040. Sixteen countries have established goals or resolutions to phase out internal combustion engine vehicles.[25] It should be noted, however, that no government has actually enacted legislation to achieve those goals.

The transition to electric vehicles takes place as people purchase new vehicles. As this book goes to press, the price of oil has collapsed, used-car prices are dropping, and unemployment has soared to levels not seen since the Great Depression. Widespread work-from-home policies change the economic model of individual car ownership. Cheap gas, cheap cars, and high unemployment will dramatically lower the expectations for multi-passenger EV sales in 2020, though smaller electric vehicles such as e-bikes might grow rapidly in popularity. The trend will resume, for all the reasons previously stated, but the pace of the transition to EVs will definitely be slowed in the near term.

## Automation

The emergence of the driverless car, though a seemingly instantaneous development, is really the result of a piecemeal evolution of autonomous technologies developed over the past twenty years. Systems like antilock braking, cruise control, antiroll technologies, light-adjusting headlights, stability control, steering assist, collision avoidance, and parallel-parking assist were all developed independently. But together with intelligent software systems and new sensor systems, cars now have the ability to deliver passengers from point A to point B without the need of a driver.

As of 2019, 10 different automakers, partnering with a variety of software companies, are investing hundreds of millions of dollars to put fully autonomous vehicles on the road.[26] Google is testing out the first commercial autonomous taxi service.[27] Uber plans to deploy fleets of driverless cars.[28] Several auto manufacturers are teaming with ride-sharing services for autonomous vehicles.

Autonomous vehicles (AV) are seen as a solution for reducing traffic casualties. Since more than 90 percent of serious auto accidents are caused by the drivers, the AV movement makes a strong case that filling our roads with AVs would be safer.[29]

The disabled community is also a strong set of customers for the new service.[30] They rightly see AVs as giving them more freedom to move about their lives.

Many cities have plans to allow customers to use AV services in the near future. It is clear that automakers and government regulators see both a public benefit and a market for this technology. Everyone wants to roll it out as soon as possible.

Rocky Mountain Institute's "Peak Car Ownership Report" paints a fairly optimistic picture of the impact of electric AVs, and the rapidity of the transition away from privately owned, fossil-fueled cars and trucks. According to the report, full deployment of an electric automated vehicle system would lower the cost to the customer by half, dramatically reduce carbon emissions, and virtually eliminate accidents. Finally, they think the increase in electric demand can be limited to 10 percent or even lower with good planning and control mechanisms.[31]

The major advantage of a fully autonomous transportation system is that the total number of vehicles on the road will be reduced. The idea is that if cars were not parked, but used constantly, the total number of vehicles needed to serve the population would be less.

Of course, the problem is the well-known "rebound effect," where we start to consume more if it is made easier and cheaper for us. Some call this the "hell" scenario, where everyone and everything is on the road because it is so convenient, and we have the same congestion. It might be a cleaner and quieter rush hour, but the congestion would be the same. Some have suggested that congestion pricing is a solution to this problem, because the ban would be on vehicles—including autonomous electric vehicles.

While there are many advantages to autonomous vehicles, there are certainly obstacles to their full deployment. The history of AI has been one of starts and stops, with some development plans experiencing an "AI winter." Self-driving cars may face problems that more data and deep learning programs will not be able to solve.[32] Even common problems like snow are still major obstacles. Automakers are starting to acknowledge that they have been optimistic in solving the AI problems.[33]

One of the problems with autonomous vehicles are human drivers in other vehicles. Anticipating what other people are going to do, whether in a car, walking, or on a bicycle, is turning out to be the most difficult obstacle. Yet many of the advantages of the AV scenario depend on near total AV saturation of our highways. If major impediments to AV are human drivers, how do we get through the transition? Are we going to ban human drivers or pedestrians on the street in order to make the system safer? Designate AV-only zones? What if car lovers, pedestrians, bike riders, and scooter riders revolt?

And the most fundamental issue may be whether people want to ride in a car that drives itself without the assurance that they can take control of it.

AV technology will probably move faster than regulations. There will almost certainly be self-driving vehicles ready for the road, in the opinion of the automakers, before state and federal agencies get rules in place to govern them. The first autonomous vehicles on the road may be limited in their areas of operation. May Mobility is already operating autonomous shuttles in Detroit, Michigan; Providence, Rhode Island; and Columbus, Ohio. But these are six-passenger golf carts traveling defined routes and limited to 25 miles per hour.[34] There are numerous public situations to be considered, and public acceptance may be easier to gain first with dedicated routes, or other limited and controlled situations. Fixed delivery routes may be some of the easiest to manage for autonomous vehicles; Volvo sold its first commercial autonomous truck to a mining company to run a fixed route from the mine to the port.[35]

Another possibility is to have a remote driver either operating the vehicle, or ready to take control. The first driverless cars are almost certainly going to have remote teleoperators either driving the car or standing by. California law requires that the vehicles must allow for remote control if they do not have a driver inside.[36]

Every major automaker and player in the AV field is developing systems to control their vehicles from afar, which makes sense if you consider the valuable robot they are putting on the road. Start-up companies like Designated Driver and Phantom Auto are offering teleoperation services, and teleoperations is likely to become a major service sector.[37]

We may see the first widespread use of autonomous vehicle technology in the form of truck platooning. Platooning means trucks are linked electronically, and follow a lead driver on highways. The trucks can be fully autonomous or have drivers who join and depart platoons on the way to their final destination. Trucks would join the road train on the major highways and drivers would be able to read, eat, or do whatever they wished in the vehicle as long as it was electronically hooked into the caravan. Because they are electronically linked, they can follow each other at a distance of only a meter apart, thus reducing air resistance like a railroad car does, saving fuel. There are also safety features on the trucks that should reduce accidents. As they approach their exit, drivers would electronically detach from the caravan and resume driving control of the vehicle.

Europe is developing a highway platooning system that should see the first platoon commercial operations by 2023. Highway tests have also been conducted in California, Virginia, and Japan. Daimler-Benz is developing the technology on public roads in the U.S. with Freightliner.[38] Tesla is working on a similar development, and Peloton Technology is also working with other automakers on its own platooning or road-train system.[39]

Platooning is likely to become a major form of freight transport on our highways. It is likely that autonomous truck platoons will hit the road ahead of robo-taxis in the cities. There are fewer variables to conquer, and the movement of freight does not have the same social obstacles as riding in a driverless car does.

While there was initially some fuss by commentators that automated trucking would eliminate jobs for truck drivers, rather it is the other way around: the shortage of truck drivers might help facilitate the adoption of self-driving trucks.[40] Indeed, even the

American Trucking Association is supporting the use of platooning as a response to a severe shortage of truck drivers.[41]

## Efficiency

In the first chapter we discussed how future systems like transportation would meet their functions with less material and less motion. This increase in efficiency is another way we are cutting our oil consumption. In the U.S., Corporate Average Fuel Economy, or CAFE standards, have gradually increased the efficiency of the engine and other characteristics of the car. Almost all motorized countries now have some sort of efficiency standard in place.

In transportation, vehicle efficiency (outside engine or motor improvements) can be gained by 1) lighter materials, 2) less motion, and 3) reduced resistance to motion.

The strength-to-weight ratio of carbon materials is far superior to steel, aluminum, and other metal alloys. Cars and trucks are starting to incorporate carbon into their vehicles, as are other transportation modes.

The aviation sector naturally focuses on lighter materials. Boeing and Airbus are designing passenger aircraft with external skins constructed almost entirely of carbon fiber. And everything on board is "lightweighted," down to the materials used in the tray carts. There is GPS-guided ground taxiing, and the takeoff and descent trajectories are smoothed out to save fuel. Aviation even has its own version of shape-shifters, as parts of the wing are being planned with material that changes its shape during flight.

Railroads, for their part, can capitalize on their already-existing advantages. Aerodynamically, trains have always had an efficiency edge over individual vehicles like trucks. Unlike independent vehicles, each of which has to punch a hole in the atmosphere and expend more energy doing so, train engines only have to punch a single hole, allowing the cars that follow to slip through behind them.[41]

And it's not just aerodynamics. Trains' steel-on-steel, wheel-to-rail interface has a much lower coefficient of friction than a truck's rubber tires on pavement. A locomotive can move a ton of freight using on average a gallon of diesel per 473 miles traveled; in general, rail transportation is four times as efficient as truck transportation.[42]

Programs that govern the speed and spacing of the locomotives for safety and fuel savings and maximize railyard capacity now direct rail systems. And stronger freight cars carry more weight with less energy.

Maritime shipping is also focusing on efficiency. Through improved fluid dynamics, maritime manufacturers are creating hulls and propellers that incur less drag. And they are incorporating sails, wind rotors, and solar panels. These efficiency gains are especially important for the parts of our transportation system that cannot be easily electrified.

## ALTERNATIVE FUELS

### Biofuels

The U.S. has focused on corn-based ethanol. Policies here subsidize the farmers growing it, and mandate the refiners to use it. Gasoline in the U.S. is blended with 10 percent ethanol. Biofuel, which is mostly ethanol, made up 5 percent of the total transportation energy consumption in 2018.[43] Although ethanol was started as an environmental alternative to petroleum, it is turning out not to have a great carbon footprint when everything is accounted for, especially indirect land use impacts. Plus, it can compete with food crops, raising moral dilemmas.[44]

The hope was that corn-based ethanol would eventually be replaced with cellulosic ethanol, made from agricultural wastes that were not being used for food or animal feed. It was assumed that the carbon footprint would at least be neutral, but the process to turn this into fuel has been more difficult and expensive than anticipated. Production has fallen well short of government mandates signed into law by President Bush.[45]

In Europe, where biodiesel from rapeseed oil, palm oil, and recycled cooking oil is more prevalent, biodiesel and a small amount of biogasoline made up 3.5 percent of the transportation fuel use in 2016.[46] Much of the European biodiesel comes from palm oil plantations. The destruction of the rainforests in Indonesia for these plantations can be vividly seen as smoke covers much of the South China Sea each year from the burning. Orangutan habitat and critical rainforest are being devastated for European biodiesel.

On the other hand, Brazil has seemed to find a way to sustainably grow biofuels and integrate them into their transportation. Brazil uses sugar cane, which is not a vital food source and the plantations are not displacing old habitat. Some of the Brazilian sugar plantations have sustainably farmed the same land for centuries.

In fact, Brazil's large sugar cane resources and ethanol industry make it the second largest producer of ethanol and largest exporter of ethanol in the world. With such a developed ethanol industry, and one that is arguably more sustainable than the corn-based U.S. industry, the fuel ethanol content in the gasoline in Brazil is 27 percent, much higher than the standard 10 percent in the U.S.[47]

And Brazil has worked on making sure the vehicles are ready. Almost all new cars in Brazil can burn either ethanol or gasoline, so people can choose at the pump whichever is cheaper. Brazil is a good example of biofuels replacing petroleum for conventional vehicles.

Algae has also been promoted as a future biofuel. Algae has many traits that make it desirable as a prospective source of sustainable energy. Most of its mass can be converted to energy, either through burning to get heat or power, or by extracting the lipids to make liquid fuels. It can be grown in saline water or wastewater, and like all photosynthetic fuel feedstock, it pulls $CO_2$ from the atmosphere.[48] We know algae can be used to make liquid fuels, as it is a key source of conventional crude oil. The trick is to manage that conversion over days using industrial processes rather than over millions of years using geological processes. Unfortunately, fuel conversion today requires more energy than the crop yields, and algae fuel has turned out to be difficult to grow and harvest in large volumes. With its high protein content, algae might be more likely to power your workouts rather than your vehicles for the near-term.[49]

### Natural Gas

Compared with petroleum, natural gas also has some benefits as a transportation fuel. It can be cheaper, the U.S. has vast stores of it, and its emissions of $CO_2$ are lower than gasoline or diesel. The delivery service UPS added more than 700 compressed natural gas (CNG) vehicles to their fleet as part of their long-term goal of reducing their $CO_2$ emissions.[50]

Public transportation is another sector turning to natural gas as it decarbonizes. As of 2019, 29 percent of buses in the U.S. were fueled by CNG.[51] While this might be a cost-effective and relatively simple transition for vehicle fleets with centralized fueling locations, it doesn't seem likely that we'll see large numbers of personal vehicles switching to the lower carbon fuel, especially given the increasing availability of electric vehicles. Honda had a Civic that ran on CNG and Ford made a CNG version of their F-150, but neither model has been in production for the last few years. For customers in India, Ford introduced a CNG version of the Aspire in late 2018.[52]

Again, there is a fueling infrastructure problem. Even though natural gas vehicles have been around for decades, and many homes have natural gas plumbing, the fueling stations still number less than a thousand. And although natural gas may be cleaner than gasoline at the point of combustion, it is still a fossil fuel that should ultimately not be combusted for the purpose of transportation.

## OBSTACLES

There are also several reasons that the transition to a cleaner, more efficient transportation system will be slower and more expensive than we want.

First we should remember that the low cost of gasoline and diesel and other petroleum products is going to remain a primary obstacle to the adoption of alternative fuels. The recent collapse in oil prices reveals an abundance of oil supply, and the oil and gas industry will be able to supply fuel at very competitive prices for decades. Due to the recession, it may be several years before oil demand returns to 2019 levels, and prices should remain low.

There are also massive sunk investments in petroleum and its manufacturing and distribution infrastructure. The fossil fuel infrastructure represents trillions of dollars of steel in the ground. This investment can't be quickly paid off or altered without major economic impact.

Some of the world's largest companies, the fossil fuel giants, will see their value reduced by half or more if the world community decides that we simply cannot burn all the oil, gas, and coal currently in the ground. The devaluation may have begun in earnest this year with the collapse in oil prices and the even more stark collapse in market valuations of fossil fuel companies.

Also, there are extensive and expensive new infrastructure requirements for the transition to renewable energy. Investment in renewable energy has been strong, but there is still a very large amount of renewable energy capacity, energy storage, grid upgrades, and charging stations needed to replace all the cars on the road with EVs.

Those who invest in transportation-related renewable energy resources such as biofuels or hydrogen still generally expect a reasonable rate of return within a few years, and it is difficult to achieve that in competition with cheap gasoline.

There are also issues with the availability of skilled labor needed to make significant changes in the fuel landscape. The existing labor force in the fossil fuel industry is relatively old, and retirements and the loss of knowledge have become a serious problem. The lack of a skilled workforce to build, install, operate, and maintain a new renewable energy infrastructure is also daunting.

Finally, there is vehicle turnover time. When the price of an EV first becomes the same as a conventional car, they may quickly dominate new car sales. However, people who have an operational car are not going to suddenly buy a new or used EV. The turnover rate for vehicles now is almost 12 years.[53] So even if EV achieves price parity and is unquestionably cheaper to operate, it will still take several more years to get gasoline-powered cars off the road.

## DIFFICULT TRANSPORTATION SECTORS

The movement of freight, and the aviation and maritime sectors will be more difficult to wean from petroleum than light-duty cars. Most of the changes in the car culture and technology that are eliminating oil from our cars and trucks are not taking place in the movement of freight. While the use of gasoline and diesel for our passenger vehicles is in decline, the use of petroleum for freight delivery is increasing.

First, the fuel-efficiency improvements for these heavy-duty vehicles are not as great as cars. While passenger cars are anticipated to gain 50 percent in fuel efficiency by 2040, the projected CAFE increase in heavy-duty vehicle energy efficiency is tiny. The

2012 rate of 6.7 mpg for heavy duty vehicles (HDV) is only expected to reach 7.8 mpg by 2040.[54]

Also, the alternatives to the car encouraged by smart cities—bicycles, walking paths, transit-oriented development, mass transit, electric two-wheelers—do not provide an alternative for freight delivery.

Freight consumption may accelerate even more quickly due to internet shopping. Anyone who observes daily fleets of UPS, FedEx and Amazon trucks zipping around their neighborhood is seeing the power of near-instant gratification at work. As people continue to buy goods that "magically" appear shortly afterward on their doorsteps, the increase in petroleum-based fuel use by freight will continue to grow.[55]

Almost all of our freight is moved around the world by two technologies: the jet turbine and the diesel engine. More than 95 percent of U.S. freight movement is powered by the diesel engine.[56]

Both the diesel engine for long-haul freight and the jet turbine are difficult to replace with batteries and electric motors. Alternative liquid fuels are available, but typically only at blended levels and even then can face performance challenges at really high or low ambient temperatures. Today they are also more expensive than the kerosene or bunker fuel currently used.

Automation and electrification of the "last mile" delivery to the doorstep will certainly reduce oil consumption, but full deployment will take time, especially in developing countries. So, in conclusion, we would expect oil consumption from freight delivery to continue to grow globally for some time.

### Aviation

Modern aviation is dependent on the jet turbine and the internal combustion engine. We will see in the next chapter that short trips can be battery powered, and batteries can even assist in regional travel. We will have electric air taxis. But battery-powered aircraft cannot handle trips beyond 500 miles at this time.[57]

We simply do not see batteries developing to the point that they could lift a hundred people 30,000 feet into the air and fly across the ocean or a continent. Battery technology would have to improve by orders of magnitude, and there is no indication that we will be able to achieve that, although we would not declare it impossible. We just rate it as highly unlikely, particularly since there are other alternatives that would be easier and cheaper to achieve and still be environmentally sound. Instead, we will need to develop the best synthetic or biofuel substitute we can for jet turbines and engines and learn new tricks to save fuel.

The aviation industry is looking toward biofuels to reduce both their dependence on traditional jet fuels as well as their carbon footprint. Airlines are also projected to move up to 7.2 billion passengers by 2035, but make as little as $10 of profit on some tickets.[58, 59] Fuel is the highest operating expense for airlines, and the price of jet fuel is often volatile because it is tied to the price of crude oil.[60, 61] Additionally, aviation accounts for 2 percent of global, human-caused, greenhouse gas emissions, and could grow to 3 percent by 2050 if left unchecked.[62, 63] In 2009, the aviation industry agreed to reduce emissions 50 percent by 2050, relative to 2005 levels.[64] Member countries of the International Civil Aviation Organization have also agreed to target carbon-neutral growth.[65] While biofuels in general are not new, improvements are required for their use in jet engines.

There are three main considerations for successful synthetic and bio-jet fuels: safety, environmental benefit, and commercialization. While aviation biofuels meet the safety and quality standards, they have yet to overcome problems with commercialization. A 2017 report from NREL found that it would be possible for biofuels to supply 30 percent of jet fuel demand by 2030, but only given an aggressive policy and/or investment scenario.[66]

Moreover, biojet fuel is still in the experimental phase, mostly for demonstration purposes.[67] Some of the most promising test results come from Neste Corp., a company based in Finland. Neste has created a drop-in biojet fuel from nonedible vegetable and animal wastes and residues, which has been successfully tested in commercial flights. Neste also has the advantage of already operating two biodiesel refineries that could be shifted to produce their biojet fuel at scale, if their fuel is approved for global use.[68]

Synthetic fuels are those that can be manufactured from renewable electricity and inputs such as hydrogen, pure water, and carbon dioxide. Converting hydrogen into liquid form can help achieve the energy density required for transoceanic travel. This pathway to producing fuels overcomes some of the limits of biofuels, and works well with existing engines, but at this time remains inefficient and costly.

## Maritime

Maritime shipping is a particularly difficult sector to transfer to renewable energy. Hydrogen, biofuels, and batteries have all been explored as alternatives to the very polluting bunker fuel currently used. Biofuels can provide a 50/50 mix with diesel fuels, but biofuel supply is limited and the low fuel density in comparison to diesel makes this alternative much more expensive.

Maritime shipping will be hard to electrify. Hydrogen and batteries seemed limited to short, regular trips between ports or for use as auxiliary power while in port.

Norway, Denmark, and Sweden have been launching electric and hybrid ferries since 2015,[69] San Francisco debuted its first hybrid ferry in 2018,[70] and Seattle, which boasts the country's largest ferry fleet, plans to start electrifying in 2021. As far as large vehicles go, ferries are an easy target for electrification given the short distances they travel and lengthy docking between trips, and this trend has the potential to not only benefit harbor air quality, but also the reduction of noise pollution will be good for marine mammals and surely appreciated by humans too.[71]

Norway launched the first electric-powered, zero-emissions container ship. However, as an article in *Spectrum* noted, "Today's state-of-the-art diesel container vessels...carry 150 times as many boxes over distances 400 times as long at speeds 3 to 4 times as fast as the pioneering electric ship can handle."[72]

The most sustainable mode for international shipping is returning to the wind and ocean currents. The trade-off here is time and volume. The clipper ships of old certainly did not add to greenhouse gases in the atmosphere or add much pollution to the oceans, but their speed and direction were limited to the wind and ocean currents, and not the direct route between ports of current cargo ships. And the volume of freight that can be transported is much less than the large cargo carriers of today. If we, indeed, want to transition international shipping away from fossil fuels—and not focus on nuclear-powered vessels—then slower, wind-powered sailboats might play a growing role.

However, we are not restricted to Victorian-era sailing technology. So-called SkySails and Flettner Rotors are just two prospective wind-harnessing designs that could make tomorrow's ships 20 to 30 percent more efficient.[73] It's easy to imagine these combined efficiency gains creating a shipping fleet that, in the coming years, is more than twice as efficient as today's.

Eco Marine Power is testing rigid sails embedded with solar panels called EnergySails. They can be stowed during rough weather. They would be combined with solar panels mounted on the hatch covers. The company calls it an "advanced integrated system of rigid sails, marine-grade solar panels, energy storage modules, and marine computers." They are conducting a feasibility study involving large bulk carrier ships.[74]

Of course, nuclear ships quickly plow through the oceans today for the U.S. Navy. But nuclear-powered ships are an expensive replacement for the big diesel engines pushing container ships today, and it is an unsettling thought to consider commercial fleets carrying nuclear materials globally out of view of international watchdog agencies tasked with fighting weapons proliferation.

Though liquid natural gas (LNG) requires two to three times the storage space of diesel fuel, because of its better efficiency, many ship lines are beginning to make the shift. In 2018 there were 77 LNG ships in operation and another 79 on order and the number could soon reach more than 200.[75] Another appealing option is to use ammonia as a maritime fuel. Ammonia stores in liquid form at normal temperatures and pressures and is a dense hydrogen carrier that does not emit $CO_2$ when reacted.

As with aviation, another idea for maritime fuel needs is to combine hydrogen and atmospheric carbon to make synthetic hydrocarbons that could be burned in the diesel engines. That would at least make the fuel carbon neutral. The challenge is producing fuel from atmospheric carbon economically.

## CHAPTER 5 SUMMARY

1.  Oil demand should peak in the coming decades as transportation technology moves away from petroleum.

2.  Cars, vans, and trucks should all be electrified within the next few decades.

3.  Self-driving vehicles are ready now, but will be first used on fixed routes and for freight delivery.

4.  Long distance travel by air and sea will be difficult to wean from petroleum.

*Mary was running late. Yet again. Her ocular lens implant flashed a reminder that her flight was leaving in 90 minutes, which seemed like some kind of cybernetic rebuke. The heads-up display on her bathroom mirror showed the air taxi en route by the most efficient path, but she really wasn't ready to rush up to the roof. AI sucks, she thought, as she lifted off in the robotic drone to the spaceport.*

*In the departure lounge, she grabbed a cup of coffee and a donut on the run. In a previous incarnation this had been an airport, and a big one. But the ubiquity of commercial suborbital flight had made the term "airport" archaic. Now she could make the New York-Tokyo flight in under two hours. The only drawback was that the vertigo and dislocation of rocket lag made her grandfather's jet lag seem like a day at the beach.*

*She left a bright morning in New York but stepped out into the middle of the night in Tokyo. She could look up from their spaceport and see the beautiful geometric patterns of the cargo drones filling the air at the nearby freight port. They moved with eerie monotony, guided by an unseen AI dockmaster.*

*Mary rolled her bag onto a hyperloop pod to speed her to Kyoto. With a whoosh, accompanied by a slight pressure on her inner ears, they were off.*

*After a retina scan confirmed her reservation at her downtown hotel, Mary entered her room, unpacked, and set up her workstation. Although she was tired, she still needed to get some work done, and the trip had already taken up most of the day.*

# SENTIENT-APPEARING TRANSPORTATION

Science fiction has given us amazing images of future vehicles. The TV show *Knight Rider* featured KITT (Knight Industries Two Thousand), a car that seemed to have a mind of its own.[1] The *Transformer* movies left us wanting cars that can convert themselves into giant fighting machines, and of course, we have been waiting for our jetpacks or flying cars ever since *Buck Rogers* or *The Jetsons* aired.[2] Future transportation systems may, indeed, have some of those features. We will have flying taxis, but they won't be cars. And automobiles are starting to talk to us, but they do not appear to have minds of their own, yet.

In this chapter we will explore some of the ways we will travel, and how we will interact with an intelligent transportation system managing the movement of people and goods. Automation will be pervasive, and flying from one part of the world to another will become much faster for the first time in decades. Aviation will be powered more by electric motors for short flights, and supersonic planes are returning. And we are starting to move into space tourism. The hyperloop could speed travel on the ground or underground, and gondolas may offer a more gentle and scenic aboveground option. Automated "ghost ships" will plow the seas.

But let's start with the way we will relate to and deal with the transportation systems of tomorrow. As with buildings, we will be talking to our vehicles and interacting with the world through them. And the vehicles will be talking to just about everything.

## SENTIENT-APPEARING TRANSPORTATION SYSTEMS

At first the thought of a transportation system that talks to us and is seemingly autonomous while we are inside can seem somewhat scary. But it makes more sense when you think of this future system as an expansion of the apps, services, and transportation vehicles that we are already using.

Today we get a reminder on our phone about our next meeting. We then are informed of the quickest route to the appointment based on traffic congestion. Travel apps help us plan trips, give us regular updates on the status of our flights, find hotel deals, and locate restaurants along the route or near our destination according to type of food and price. Also today, when we receive an email referencing a future date or time, it is highlighted and a click automatically pulls up our calendar to add the event.

If we have a phone that synchronizes with a car system such as OnStar, we already have the option to push buttons on the steering wheel or use audible commands to make and answer calls, play music from our stored library, or ask questions of our digital assistant. If we are not sure of a location, a GPS voice from our car speaker will guide us turn by turn to the building.

Now consider more advanced travel systems of the future that will sync and coordinate your schedule, personal travel apps, vehicle systems, traffic control schemes, and artificial intelligence related to your trip.

Travel apps and personal digital assistants will become our travel agents, authorized to plan and purchase flights, hotel rooms, and make dinner reservations. The AI system will know your schedule and your preferences, such as aisle seats and Thai food. It will also be monitoring everyone else's preferences and changes in travel plans, which will give you more flexibility if you need to make a change to find a cheaper fare or better route. Artificial intelligence systems will listen to our conversations and develop travel plans on the fly.

This could be very useful in meetings. As soon as you end the meeting, your travel app could show you options for the cheapest, fastest, or most environmentally friendly choice to get to that new appointment next month. And since you have authorized the app as your travel avatar, all the bookings are made with a voice command. You choose something, and the system will have the appropriate modes of transportation ready for you at the appointed time—all the way to and from the destination.

Your future travel app may also communicate regularly with other AI systems to make your trip more convenient, personalized, and safe. For instance, the app may have received a communication from your doctor's AI that you have broken a bone in your right foot and travel plans must accommodate a cast.

When you get into a car, it may, indeed, greet you by name and inform you about traffic or present entertainment options for your trip. Future vehicles will have access to the same technology that buildings support today with facial-recognition software and other sensors and programs for identification.

With advanced technology, the car may be able to turn the inside of the vehicle into a holographic entertainment platform, which passengers will be free to enjoy without concern for driving. Not only that, but if we were also able to tune in to the communications going on between the vehicle and the world, we would suddenly realize how talkative our future vehicles have become.

Today, traffic engineers talk about "V2X," which stands for "vehicle to everything." It is the expectation that vehicles of the future will be "talking" to just about everything.

Here's a partial list:

- *V2I—Vehicle-to-Infrastructure.* The car is going to be talking to traffic lights and crosswalks. Parking spaces may be broadcasting their availability and guiding cars to them. Drivers may get information on traffic and weather conditions. AI and remote operators may also control your car when you are not there. The car may drop you off at the restaurant door and then find the nearest parking space or move to another user or find a place to charge while you eat.

- *V2V—Vehicle-to-Vehicle.* Vehicles are going to be talking to each other to avoid collisions and integrate into traffic. They may search for a platoon of vehicles going in the same direction, or to the same destination. V2V will allow the vehicles to flow in traffic with minimal space between them, saving energy.[3] Vehicle information sharing and coordination should result in the near elimination of accidents, reduced travel times, and the disappearance of a thing once known as the traffic jam.[4]

- *V2G—Vehicle-to-Grid.* The car will be exchanging information with the power grid to determine the best time and location to recharge the battery, determined by cost, strain on the grid, and availability of charging stations. Vehicles will also be able to discharge power to the grid and assist in grid stability and support, and the utility will purchase the power.

- *V2H—Vehicle-to-Home.* One of our colleagues, Dave Tuttle, wrote part of his PhD thesis on the vehicle-to-home connections that would enable vehicles to power a home during grid outages. In Japan, Nissan passed out power cords after their nuclear accident to allow homeowners to get power from their cars. And there is a utility truck on the market that will power homes while repairs are being made to the grid.

- *V2P—Vehicle-to-Pedestrian.* Communicating to pedestrians for safety from the approaching vehicle.

The sophisticated communication and sensing array in a future vehicle will also be providing real-time and recorded history of events. In a criminal investigation, the video from cameras mounted on nearby buildings is often checked. Future investigations may start pulling additional information from nearby parked or passing cars.

There will be different levels of AI in the transportation system, and you will only interact with some of them, like booking agents, and car systems. AI traffic-control systems will control the vehicles, including yours, as you move about your day. The vision traffic engineers have of a rapid, smoothly flowing city of vehicles will only work if you give up control of your car.

Which brings us to an uncomfortable discussion. We will be moving about inside robots. We may cringe at the thought of being inside a robot, but that is because of our image of a robot. In fact, we already are transported from place to place inside steel boxes not in our control, and consider it quite normal. Whenever we step into an elevator, we are whisked up in the air in an electronically controlled steel container. But it is a short trip and we have gotten used to it as normal. Nonetheless, for anyone who has been trapped in an elevator when the electricity went out, the reality of your situation becomes terrifying.

We also have become used to being moved from one airport terminal to another on driverless train pods. And major cities move thousands of people every day with driverless subway and light rail systems.

But we are not used to losing control of our car. We are not used to our car being able to go wherever an automated driving system, either onboard or remotely controlled, tells the vehicle to go. And in that sense, suddenly the car seems to have a mind of its own.

But we are starting to adjust to that, as well. In fact, one problem emerging is drivers falling asleep or otherwise no longer engaged with the car. Yet most people are still not ready to consider boarding a pilotless airplane.

## AUTONOMOUS FREIGHT DELIVERY

The first general use of driverless vehicles will probably be in freight delivery. There are not the safety issues of protecting a passenger, and dedicated routes for freight make this an easier transport sector to automate. Driverless technology is advancing in all areas of freight delivery: air, ship, train, and vehicles.

Coordinated transportation systems will ensure that single products and bulk goods can be packaged, routed, and shipped in the most efficient fashion possible, without ever being touched by human hands. After a customer decides on the product they want and the purchase is complete, robots in an automated warehouse will locate, pick, and package the product, and place it in the proper location for shipment. The warehouse will load the product onto a self-driving truck that will deliver the goods to either an airport, rail yard, or ocean port.

### Ports

Say the package ends up at an ocean port. We now are seeing the automation of those ports, where the gantry cranes unloading ships, stacking cranes, and port transport vehicles are being automated. The port of Rotterdam is now almost fully automated and run by software.[5] We can expect similar automation loading freight at rail yards and airports.

### Cargo Drones

The autonomous truck might also deliver the package to a freight airport, where an unmanned cargo drone would fly it to the next stop. Several aircraft manufacturers and delivery services are working on unmanned aerial vehicles (UAV)—drones—to take a significant market share of airfreight delivery.

Boeing is developing a helicopter drone expected to haul 500 pounds for several hundred miles. This will be a truly autonomous drone, without a remote pilot, using software to fly it. The AT200 in China can fly 2,000 km with a cargo capacity of 1.5 tons. It is remotely piloted, but if communication is lost, "it is programmed to return to its point of origin and land by itself."

Currently, the world's largest commercial cargo drone in use is the FH98 in China. It has a cargo capacity of 7,000 pounds and a maximum range of 1,200 km. It is essentially a small cargo plane that has been retrofitted to operate like a drone.[6]

The major delivery companies are naturally working hard to make drone delivery a reality. Deutsche Post DHL, the world's largest logistics company, has begun its first regular drone delivery service in an urban setting, in Dongguan, China.[7] Amazon made the first "Prime Air" delivery by drone in 2016. They are testing drones in several countries and working through the testing and regulatory process. Their goal for Prime Air is delivery of packages of up to 5 pounds within 30 minutes.

Start-ups Volans-i, Natilus, Matternet, and Zipline are trying to carve out niche segments of the freight business for their autonomous flying drones.[8] Volans-i, for instance, has successfully developed a model that can take off or land on any flat 15 × 15 ft. platform, has a 500-mile range, and can carry 20 pounds. If the load-bearing properties of this drone are improved, this technology could allow remote operations, even those at sea, to order up heavy equipment on demand.[9]

### Ghost Ships

Maritime shipping is also turning to automation, with the possibility of unmanned "ghost ships" on the open seas. Rolls Royce has started the Advanced Autonomous Waterborne Applications project in Finland to achieve fully autonomous shipping. Google is working with Rolls Royce on the AI system that will help the ship identify and track objects encountered at sea.[10]

The European Union's Maritime Unmanned Navigation through Intelligence in Networks (MUNIN) is assessing the technical, economic, and legal aspects of such "ghost ships." These ships might also be less attractive for pirates as their navigation systems could be designed to make them harder to hijack.[11]

### Last-Mile Delivery

There is also a focus on automating "the last mile" of package delivery. Ford recently announced its plans to develop a two-legged delivery robot named Digit that folds up to fit in the back of an automated delivery vehicle. The company's announcement boasts that the robot will be able to navigate around the unpredictable obstacles of a residential area while carrying up to 40 pounds.[12] This approach would combine self-driving delivery vehicles with the robotics necessary to get from the street or driveway to the front doorstep.[13]

Another option for the last mile of delivery is small robotic devices that would travel on sidewalks up to the front door. These would be slow-moving, driverless electric pods delivering packages to homes and offices. The customer could open the small robot pod with their smartphone and retrieve the package, or arrangements could be made for the building to receive it.

More orders will be delivered beyond the doorstep and put in their proper place—such as the refrigerator, pantry, or trunk of your car. Amazon has tested such a car trunk delivery program. And earlier in this book, we saw how buildings will be able to accept deliveries.

## Autonomous Fueling

Another feature of the autonomous transportation system will be fueling. Electric vehicles could go to robotic charging stations or park over inductive charging pads.[14] Drones and maritime vessels could also be fueled automatically. The first autonomous aerial refueling of a drone was accomplished in 2015.[15]

The Naval Surface Warfare Center has developed an automated, floating fuel station that can be airdropped to provide fuel for autonomous ocean-going vessels, extending their range.[16] There is even a robotic gas station addition called Autofuel, which can be added to conventional gas stations to refuel gas-powered cars.[17]

Overall, autonomous freight movement systems will be able to offer options for speed, cost, and environmental impact. Sustainable shipping companies may offer slow shipment by wind- and solar-powered clipper ships. On the other hand, supersonic aircraft could deliver emergency supplies immediately, since cost constraints would be less important.

## ELECTRIC AIRPLANES

The shift from internal combustion engines to electric motors in the automotive industry is also changing part of aviation. Carbon fiber materials, advancements in electric motors and batteries, and advanced navigation systems have enabled airplane manufacturers to develop electric prototypes. The electrification of aviation can be viewed in terms of three ranges of distance: short hops, regional travel, and long-distance flights.

Several companies are currently testing prototypes of electric or hybrid-electric aircraft that could very well start operating as flying taxis. These aircraft are eVTOL (electric vertical takeoff and landing): they take off and land vertically, like helicopters, but unlike helicopters, they have wings that allow for much more efficient forward motion. Another advantage over helicopters is that these aircraft, at least the electric models, are quieter.

Aurora Flight Sciences, a company owned by Boeing, is a partner in Uber's Elevate program, which has a goal of air taxis operating in Dallas and Los Angeles by 2023. UberAir aims to be cost-competitive and believes it will be widely used in Los Angeles by 2028.[18] Another partner in the Elevate program is Bell, which has a prototype that could seat five and travel 150 miles. Airbus is also taking the idea of flying taxis seriously and has its own program, Vahana, in development.

Since the cost of operating an eVTOL is not much less than a helicopter, the cost of a ride may be similar, and the initial market will probably be the same people who can now afford a helicopter lift. And yes, there are flying cars in development that can both drive on roads and fly. NFT, AeroMobil, and Terrafugia are developing such dual-purpose vehicles.

Battery limitations mean that local air taxis are going to be good for short trips across the city and travel to the regional airport. They will still need to land and take off from downtown rooftops, such as the helipads we see today. However, the problems associated with putting lots of flying vehicles over our cities should be obvious.

It is interesting that Uber is advertising relieving city traffic congestion by putting "tens of thousands" into the air. Not only will air traffic congestion become a major issue, but also noise will be a concern. A new design using twin rotors turning in the same direction is supposed to make them quieter, but they are not as silent as electric cars. People already complain about the noise of a relatively small number of drones.

Regional jets could be the first aviation sector that sees widespread electrification. If you take a flight that is not cross-country a decade from now, the likelihood is nontrivial that it will be on an electric or electric-hybrid aircraft. An electric-hybrid aircraft substitutes one or more conventional jet turbines with an electric motor.

For flights fewer than 500 miles, electric aircraft can be much cheaper to operate. In a hybrid combination, you can switch on the electric element to assist in the key parts of the flight—takeoff and landing. As the technology develops, thanks in part due to electric automotive advances, the potential savings on aviation fuel costs presents a major incentive.

One advantage of the relatively quieter operation of electric and hybrid aircraft is that they are less likely to violate local noise ordinances. That means they can land and take off earlier in the morning and later into the night, when traditional planes are prohibited. Operating some flights during these off-hours will enable more throughput at the airports, relieving congestion and improving efficiency of the air traffic system.

Royce, Airbus, and Siemens are working on E-Fan X. It is a 2MW electric motor mounted on a BAE 146 jet, and expected to fly in 2021. United Technologies is also working on a hybrid electric plane. EasyJet says it will start using electric aircraft by 2027, and Zunum Aero, backed by Boeing, is also working on a hybrid.

An Israeli firm, Eviation, debuted an all-electric passenger aircraft capable of traveling 276 mph with a range of 650 miles, and the orders are already coming in, including from U.S. regional airline Cape Air, which intends to buy a "double-digit" number of the aircraft. Similar planes would use $400 in fuel for a 100-mile flight, while this one would only be $8–$12.[19] If the cost savings of electric engines are passed on to consumers, electric flight could make quick interregional travel an affordable option for many people.

Lilium Aviation claims it will make the world's first all-electric vertical takeoff and landing jet, with a range of 183 miles.[20] And Joby Aviation plans an all-electric plane with a 100- to 200-mile range that people will summon like an Uber ride.[21]

The first hybrid electric aircraft may be a retrofit. Companies like Ampaire are retrofitting two-engine planes by replacing one of the engines with a battery-powered electric motor. One advantage of this approach is that it may be quicker to get FAA approval, since new plane design approvals can take years. Replacing just one of the two engines with an electric motor can cut fuel consumption and maintenance in half.[22]

These electric and hybrid aircrafts are certainly an important technological development to reduce global carbon emissions, but unfortunately, 80 percent of the aviation industry's emissions involve ranges that for now seem entirely impossible to electrify or even hybridize.[23] Batteries can provide assistance on takeoffs and landings, and electrification at airports can power jets while on the ground, but low-carbon liquid fuels remain necessary for long-distance flights. What will be changing is speed.

## SPEED

The speed for freight might be slower for certain cargo and particular routes when airships or sailboats with Flettner rotors are used, but the speed of travel for modes of transportation carrying people is going to be increasing. We will see the return of supersonic aircraft, and the hyperloop will be the new speed limit on rail. There may be some speed increases for automated vehicles. Overall, we will move people (and most goods) faster a few decades from now.

There was a time, in the 1950s and 1960s, when supersonic passenger travel seemed exciting and imminent. But it takes about the same amount of time to fly from New York to Los Angeles today as 70 years ago. This limit, however, is not a technological one, but a practical one. When aircraft reached supersonic speeds, the resulting sonic boom broke windows and created havoc along the flight path.

Consequently, supersonic travel over land was banned in 1973.[24] This, together with a string of mishaps, like the wreck of the Soviet TU-144 "Concordski" at a Paris Air Show in 1973,[25] and the fatal Concorde crash at Charles De Gaulle Airport outside of Paris in 2000, ended passenger supersonic travel in 2003.[26]

That may change with a new design by NASA that disperses the sound waves and prevents a sonic boom. Today, companies like Boeing and Lockheed-Martin, with assistance from NASA, are studying how to alter sonic booms to make them more acceptable to the public and the FAA. For Michael, who worked in the early 1990s on NASA's National Aerospace Plane (NASP) project, the return of high-speed travel is an exciting prospect. For background, NASP was a program launched by the Reagan Administration in the 1980s to develop planes using SCRAMJET (supersonic combustion ramjet) engines that could take off from and land on horizontal runways before crossing oceans in just a few hours. The program was killed shortly after Michael worked on it as part of budget-balancing initiatives in the Clinton administration.

Boeing and Lockheed Martin are also backing start-up Boom Technology, Inc. to develop a jet that could cut the six-hour flight time between New York and Los Angeles in half while reducing the noise to the level of highway traffic.[27] And General Electric and Lockheed Martin are working with Aerion Supersonic to develop a plane to travel 1.4 times the speed of sound. They aim to operate flights between New York and Sao Paulo and London and Beijing.

Spike Aerospace, with help from Greenpoint Technologies and Siemens, is working on a Mach 1.6 model called Spike S-512 that can make the trip between Dubai and New York in half the time of today's flights. Boeing is going big. They are developing a hypersonic model that could achieve Mach 5, five times the speed of sound, and make the trip from Sydney to San Francisco in just less than two hours. Several of these companies hope to have supersonic jets in the air by 2023.[28, 29]

Along with supersonic plane travel, people could someday get from point to point on a suborbital rocket flight. It's estimated that intercontinental travel could take less than an hour in such a vehicle. Making such craft safe, reusable, and not cost-prohibitive are all problems future generations will have to solve.[30]

Elon Musk envisions such a 'city-rocket,' with the specific goal of being able to move between any two cities in the world in under an hour. A short video shows individuals taking a short ferry ride to a floating launch pad, blasting off, and reaching 17,000 mph before reentering the atmosphere and landing on another floating launch pad. Trips between metropolitan hubs such as London and Dubai would take only 29 minutes, New York to Paris just 30.[31]

Virgin Galactic founder Richard Branson is testing craft that could send space tourists into suborbital flight. Blue Origin, founded by Jeff Bezos, expects to fly people in their New Shepard suborbital vehicle.[32] These flights go up, give the passengers the experience of weightlessness and a spectacular view of earth from space, and then return to the point of origin. Getting from point to point is much harder. With suborbital flight, you are actively fighting against earth's gravity all the way. There are still a number of technical challenges to overcome before suborbital flights could whip us around the globe for an afternoon business meeting.

Apparently, Robert Heinlein predicted "the violence" of the experience pretty well in his 1982 sci-fi novel, *Friday:*

> The high-G blastoff always feels as if the cradles would rupture and spurt fluid all over the cabin. The breathless minutes in free-fall that feel as if your guts were falling out. And then reentry, and that long, long glide that beats any sky ride ever built. Presently free fall went away and we entered the incredibly thrilling sensations of hypersonic glide. The computer was doing a good job of smoothing out the violence, but you still feel the vibration in your teeth.

> We dropped through transsonic rather abruptly, then spent a long time subsonic, with the scream building up. Then we touched and the retros cut in, and shortly we stopped. We had lifted at North Island at noon Thursday, so we arrived 40 minutes later at Winnipeg the day before, in the early evening on Wednesday, 1940 hours.[33]

Your guts will be jostled, and your teeth will grind and vibrate, but the time savings may be worth it. However, there will not be any drink carts on the suborbital.

With suborbital flight, we may gain the option of getting from London to Sydney in less than an hour.[34] However, it will be very expensive, and certainly for the richest. Technological development will bring the costs down over time, but suborbital flights will not be an option for the normal traveler for decades, if not longer.

## *Hyperloop*

Back on earth, the biggest change in speed of transportation will be the hyperloop. It's not a new idea to use magnets in evacuated tunnels to transport goods and people quickly. Isambard Kingdom Brunel was a pioneering British engineer who experimented with using compressed air to transport carriages in the late 1800s.[35] In a 1972 RAND Corporation report, Robert M. Salter laid out the concept of a high-speed underground transportation system using pneumatic tunnels as a way to improve the efficiency of transportation.[36]

But in 2013, Elon Musk revealed his vision to develop the technology, calling it Hyperloop. Hyperloop vehicles are expected to reach speeds over 600 mph. The annual Hyperloop design competition hosted by Elon Musk's SpaceX company attracted more than 20 university teams in 2019.[37]

Hardt is another hyperloop company operating in the Netherlands and has a test facility at Delft, with plans for a 3-km-long test tunnel called the European Hyperloop Center. Hardt hopes to have a commercial route in operation between Frankfurt and Amsterdam by 2028, with a trip made in less than an hour.[38] Hardt Hyperloop founder, Tim Houter, says the hyperloop will be 10 times more efficient than an airplane and more efficient than trains.

Virgin Hyperloop One, formerly headed by Richard Branson, is a leader in hyperloop development and has been working with governments all over the world. Virgin Hyperloop One's Kelly says the technology will be about 5 times more efficient than short-haul flights.[39] They have a test facility in the Mojave Desert outside Las Vegas, where their test pod reached speeds of 240 mph.[40] They have also signed a deal with the Saudi Arabian government to build a 35-km test track.[41]

India is making the hyperloop a public infrastructure project. The Pune-to-Mumbai hyperloop is projected to eliminate a 3.5-hour car trip for a 35-minute hyperloop ride. Virgin Hyperloop One is a leading contender for the project.

The Pennsylvania Turnpike Commission approved a four-year contract for $2 million with AECOM to review the potential for a hyperloop system to extend across the state. The Mid-Ohio Regional Planning Commission is working with Virgin Hyperloop One to study a link between Pittsburgh, Columbus, and Chicago (48 minutes from one end to another). And a public-private group is exploring making Missouri the first location for a hyperloop track, with a 28-minute hyperloop trip from Kansas City to St. Louis.[42, 43]

Now Hyperloop One has formed a partnership with global port operator DP World to form DP World Cargospeed, envisioning a hyperloop-connected world shipping freight. Branson wrote that he wants to "deliver freight at the speed of flight and closer to the cost of trucking."[44]

Hyperloop trains could replace short-haul airplane traffic, especially if they can combine the frequency of the subway with the speed of the aircraft. Some may have concerns with the claustrophobia of such a trip, but it is essentially the same as a long subway ride, and may be worth the savings in time.

While the hyperloop concept is exciting because of its prospective benefits on operating costs, efficiency, and speed for moving goods or people, cynics note that it has all the up-front cost challenges of long-range tunneling, high-pressure pneumatic systems, and trains *combined,* thus it might have trouble competing with cheaper surface options or simply building traditional train systems. In addition, safety requirements and comfort concerns for passengers who don't like sudden acceleration or deceleration might reduce the pneumatic systems to movement of goods rather than people. However, that might not be so bad because doing so would remove freight from airplanes or surface transportation which would reduce air pollution, congestion, and traffic fatalities on roads.

## Gondolas and Airships

The future transportation options may not always mean getting somewhere faster. Many people will still want to walk, bike, or scoot around at a more leisurely pace—and they may want to take a gondola to the city center.

Gondola systems are not just for the Swiss Alps—they have been installed in several major cities such as the Mexicable in Ectepec, Mexico; Cali and Medellin in Colombia; La Paz, Bolivia; Los Angeles, CA; New York, NY; and Rio de Janeiro, Brazil. The gondola systems are envisioned to connect to bus and subway transit systems.[45]

And airships might return as a way of transporting cargo to remote areas. Sports enthusiasts are familiar with airships such as the blimps used for aerial television shots at major sporting events, but they can do more than offer a cool visual: Airships can carry huge loads and deliver them with hardly any need for infrastructure other than a landing pad, which could be as simple as a cleared field. They can haul vehicles, mining equipment, military gear, or disaster relief supplies, all while staying aloft for up to a week at a time. For point-and-drop deliveries to remote locations in mountains or jungles, an airship is much more logistically compatible than building an airstrip for planes, or even roads for jeeps in some cases.

Flying Whales, a French manufacturer, is partnering with state-owned China Aviation Industry General Aircraft, to start producing airships in 2022. Each is twice as long as the Boeing 747. A fleet of 150 machines has been ordered at a total price of 150 billion euros, and they will be built in factories in France and China. The rail-industry Alstom SA, and oil-services provider Technip SA, are both interested.[46]

Lockheed Martin is developing its hybrid-electric LMH-1 airships. Their model combines the benefits of an airship with the flexibility of a hovercraft. It is shaped like a wing, so it can combine aerodynamic and aerostatic (buoyancy) lift. The material is a tough, Kevlar-like fabric, and tiny leak-seeking robots could service this new generation of airship. With room for 19 passengers and an ability to fly 1,600 miles without refueling, Lockheed Martin's airship has attracted attention from the oil and gas industry. There is already a $500 million order from Straightline Aviation, a company that works closely with oil, gas, and mining industries. The buyer plans to lease to Arctic oil and gas companies.[47] Another project close to being in the real world is Hybrid Air Vehicles' *Airlander 10,* aka The Flying Bum. It can haul 20 tons of cargo for a week at a time.

For regulatory reasons, airships might be forced to operate out of airports, which would be a big strain on space. Airships may first be used in Canada/Alaska and Africa, assisting high-cost operations into difficult or remote territory. Airships could potentially increase the volume of cargo that moves through airports to be closer to that of what moves through seaports.[48] Or for ports such as the Ports of Long Beach and Los Angeles, which have trouble moving goods away from the ships by truck because of congestion on the highways, airships might be a way to leapfrog the highway and mountains to move freight to an intermodal transfer center further inland.

Widespread use of airships, despite their exciting potential as a low-carbon shipping method, may be stifled by the expense and availability of helium, a limited resource without any known reserves. There is a helium shortage and prices are rising, so the economics for airships may not work. Searching for oil using helium-filled airships is perhaps the height of folly.

However, hydrogen, which is even lighter than helium, might be available as a substitute. After the Hindenburg accident, the flammability of hydrogen was an international news story and was a major motivator to move toward helium instead of hydrogen gas inside the airship. However, as renewable hydrogen costs drop for other purposes, it might be a cost-competitive solution to helium shortages. The safety concerns can be more actively managed, and for autonomous airships without pilots or passengers, the risk to the freight might be worth the cost savings. In that way, airships might be the technology where hydrogen, autonomous operation, and lighter-than-air aviation converge.

## SPACE AND BEYOND

The leading entrepreneurs in the space race—Elon Musk, Richard Branson, and Jeff Bezos—are collectively known as the "space barons." Musk has SpaceX and is delivering payloads for NASA and others regularly. Branson is selling tickets for the first suborbital tourists, and Jeff Bezos's company, Blue Origin, is preparing a moon lander known as the Blue Moon.

SpaceX and Boeing have won contracts from NASA to transport crew to the space station. And SpaceX seems to have solved the reusable rocket problem. Since the rocket is about 90 percent of the cost of a space launch, the successful development of a reusable rocket by SpaceX may be an even more important space milestone than the moon landing.

Soon, our rockets could even be 3D printed. Relativity Space Inc., a start-up out of Los Angeles, thinks that their printers can cut down on the cost and time required for a rocket launch through automation; the founders believe that in four years their price for a rocket launch will be $10 million, compared to the current $100 million. They also claim that they can build an entire rocket in just a month. The company successfully test-fired one of their printed engines in June 2017 and plans to fly their first complete rocket in 2021.[49]

But space travel for the common person is not seen as part of our transportation system in the foreseeable future. Trips to the moon in the future could become like trips to Antarctica today. And there certainly could be exclusive space resorts in orbit for the wealthy and adventurous.

There could be explorers and settlers on the moon and Mars. But beyond that the question becomes, where do we go? There are not going to be faster-than-light space drives to take us to the stars. Any trip beyond our solar system is going to take generations to complete, so while space exploration relatively near to Earth may have many benefits and joys, do not plan on interstellar vacations.

Moving more equipment and people into orbit, and then on to the moon and Mars, may become much easier and less expensive if a space elevator could be built. The space elevator, first envisioned by Arthur C. Clarke, is a ribbon of ultralight, ultrastrong cable that is anchored both to the earth and to a counterweight 62,000 miles from the earth's surface. Travel to space would be accomplished by riding an elevator attached to the ribbon of carbon, which is 100 times stronger than steel, and as flexible as plastic.[50]

But there are severe technical challenges to overcome. It is extremely difficult to make a cable strong enough. Carbon nano-tubes were first thought to be a good material, but the nanotubes need to have no imperfections. Even a small kink could make them unstable. Furthermore, the longest nanotube made so far, in 2013, was about a foot and a half, a far cry from the distance needed to escape gravity. Finally, lightning strikes could disintegrate a large portion of the cable.[51]

However, a Canadian company, Thoth Technology, has secured a patent for a space elevator tower that would use a spiral elevator method to move an elevator car up to 15 km above the earth. This configuration does not attempt to solve the material problem with a cable. Rather it builds an elevator tower high enough to significantly offset the hardest part of rocket liftoff.

At the top of the tower would be a runway for space vehicles to take off and land, and allow reusable one-stage vehicles to enter orbit, potentially saving 30 percent on fuel. At the same time, the space elevator would be a wind energy generator, and a large area communication tower. They even envision a hotel on top for space tourism. Thoth plans to build a tower within the next decade.[52]

"Beam me up, Scotty." That's what Captain Kirk said when he wanted to travel from an alien planet back to the starship *Enterprise*. That sort of travel involved teleportation, an idea that most consider alien and impossible.

Architect and systems theorist Buckminster Fuller thought near the end of his life that we would soon find a way to achieve tele- portation. Many thought that Fuller had finally gotten a prediction wrong. It turns out that he might have been right, although not in the manner in which he was conceiving it.

In 1993, Charles Bennet of IBM confirmed the radical possibil- ity of something called quantum teleportation, or the transfer of information describing characteristics of photons and atoms from one place to another.[53] In 1998, scientists at Caltech proved the idea by teleporting a photon 3.28 feet.[54] By 2012, scientists in the Canary Islands successfully quantum teleported a pair of photons over 88 miles.[55]

Some scientists contend that this sort of quantum teleportation doesn't represent a Star Trek-like disassembly and reassembly type of teleportation, but merely the transfer of information at a distance. Such skeptics believe the most optimistic outcome of the technology would be transfer of information between the earth's surface and a satellite, and perhaps the eventual creation of a "satellite-based, secure, quantum internet."[56]

Others, like physicist Dave Goldberg, disagree. "Quantum tele- portation is not only a possibility—it's a reality, at least for single photon states," says Goldberg, "and there's no reason it couldn't be scaled up somewhat. I'm skeptical that it'll ever be practical to tele- port people, but there's no fundamental reason why we couldn't do so. It's just mind-bendingly complex."[57]

That complexity is related to the fact that humans are made up of more than 50 trillion units, and each of those units is much more complex than a single photon.[58] It is also hard to understand how the duplication would occur for organs inside our bodies.

Physicist Michio Kaku points out that we quantum teleported cesium atoms, and predicts that we will quantum teleport a molecule within a decade.[59] Kaku also points out that the idea of teleporting humans does, of course, raise such philosophical questions, as "is there a soul independent of the body." So even if we could perfectly transfer all the information necessary to teleport a person's body, would their consciousness and soul travel with them? Maybe we'll find out in the future.

In *The Future of the Mind,* Kaku conceives of a future with computers powerful enough to decode all the information of an individual's brain, and then transfer it to surrogate, robotic bodies that could be updated and improved ad infinitum, allowing us to travel into immortality. Kaku also claims that exploring the universe as "beings of pure energy" on beams of light is "well within the laws of physics."[60] Though Kaku admits that there are engineering difficulties, such as the advancement of quantum computers that could compute fast enough to translate a human's full consciousness, he says these problems could very possibly be solved in the next century or so, allowing us to not just transfer our consciousness to surrogates, but simply to remain, "and roam, almost ghostlike, as a form of pure energy."[61]

## CHAPTER 6 SUMMARY

1. Sentient-appearing transportation systems will assist us with all our movement through the day and across the country.

2. Giving up our personal vehicles for shared autonomous vehicles is a natural extension of being shuttled about in buses, trains, ships, and planes. It might happen sooner than we think, but won't necessarily be an easy transition.

3. The speed of transportation is increasing with autonomous vehicles, air taxis, supersonic jets, and hyperloop trains.

4. Escaping Earth's gravity requires immense quantities of energy. Extraplanetary travel will remain rare and expensive for some time.

# PART 4

# THE FUTURE OF POWER

Sam was adamant; the annual Labor Day party _would_ go on. Susan was equally forthright; her husband had lost his mind.

"What part of 'There's a storm coming' don't you understand?" she asked, and not for the first time.

"And what part of 'We've been planning this for weeks' don't _you_ under-stand?" he replied with what he thought of as a witty rejoinder. "Besides, we live in Houston. Storms come with the territory, like taco trucks and the Astros."

Twenty-four hours later, Susan was resisting the urge to shout, "I told you so!" for about the billionth time. Sure enough, the party had gone on, and sure enough, the storm had come, and sure enough, Houston's power grid had collapsed like a soggy spider web.

The partygoers were huddled in the darkened house, watching the streets outside transform themselves into rivers. Sam at least had a battery-powered radio that let him tune into the National Weather Service and the local emergency channels. As far as he could make out, the power plants were still intact, for all the good that did without a means to transmit the power to households.

The city's lighting and power company estimated it would be one to two days before power could be restored to most neighborhoods.

In the meantime, the guacamole was going south fast. "Well, at least it will make a great story," Sam said gamely. Susan just poured another glass of wine and gave him a dirty look.

# THE CHANGING POWER INDUSTRY

The electric power industry is undergoing the most dramatic change since its inception more than a century ago. The *who, what, where, when* and *why* of electricity generation is changing.

- The *who* are the owners, operators, and decision-makers of the power industry. The utility and the customer are changing roles as customers start to produce their own electricity.

- The *what* are the fuels we use to generate electricity. Decarbonization of fuels is perhaps the single biggest economic change in the industry.

- *Where* we generate electricity is also changing. Rooftop solar and other on-site generation are driving a decentralization trend.

- Energy storage is changing *when* we use electricity, decoupling the times between when we generate and consume electricity.

- And finally, new electrical workloads and power needs are changing *why* we generate electricity.

It is easier to explain the consequences of these changes if we start with *what* and end with *who*. This chapter will go through each of five areas of the changing power industry, beginning with the energy resources we are using and finishing with the changing utility business models.

## *WHAT* WE USE TO GENERATE ELECTRICITY IS CHANGING: DECARBONIZATION

If the story of humanity is a long history of harnessing energy and moving from one fuel source to another, the subtext is an inexorable shift from high-carbon fuels to ever lower ones. We began with wood and then coal as the world industrialized in the 19th century and rapidly grew into oil and natural gas in the 20th century. With each transition we have advanced to fuels with fewer carbon atoms.

# ELECTRICITY GENERATION FROM SELECTED FUELS

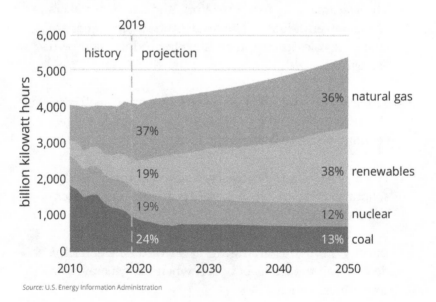

Source: U.S. Energy Information Administration

**Figure 4.**

With the move from coal to natural gas we cut the carbon content approximately in half. And as we move to more hydro, wind, solar, nuclear, and geothermal we will be tapping energy sources and fuels without carbon.

Overall, from 1995 to today, the trend in the U.S. has been toward a greater use of natural gas and renewables, at the expense of coal. For a variety of economic and policy reasons, this trend will likely continue.[1]

Things are a bit different around the world. Globally, coal power generation continues to increase year-on-year as developing countries seek to industrialize and electrify. The majority of the world's population does not enjoy the standard of living, and commensurate level of energy consumption, found in the U.S., Europe, and developed nations. Nuclear power and natural gas, too, continue to increase, especially in China and India as the world's two largest populations industrialize, electrify, and grow. Worldwide, the International Energy Agency (IEA) predicts that even in 2040, fossil fuels will still provide most of the world's energy.[2]

## Coal

Coal powers our world. It powers our buildings and appliances, and as people plug electric cars into the grid, it's powering a growing number of those as well. As mentioned, power production from coal is expected to grow worldwide before it falls. Such isn't the case in the U.S.

Domestic demand for coal has generally been in decline since 2008. Coal production saw a modest increase in 2016 and 2017, fueled by the biggest increase in foreign demand in more than a decade, but is still following an overall declining trend.[3,4] The reasons are myriad, but perhaps one of the fundamental reasons is the sheer amount of energy lost in coal power plants. Only one-third of the energy produced from the coal combustion is converted into electricity. The other two-thirds of the combustion energy from the coal is lost up the smokestack, through the heat exchanger, and via countless other small thermal and mechanical losses. This sheer waste of such a carbon-intense and finite energy resource is why many consider coal to be dirty.

But coal is also physically dirty. Consequently, environmental regulations related to acid rain precursors, particulate matter, and heavy metals have made coal use more costly. These price considerations are causing some utilities to close down older and less efficient, or "marginal," units that no longer run at full capacity. More than 250 coal plants were retired nationwide between 2010 and 2018, and EIA estimates almost 90GW more (or roughly 120 power plants) could be retired by 2030.[5]

Despite the Trump administration's efforts to roll back regulations on carbon emissions at the federal level, there are strong trends diminishing the use of coal. These drivers include state renewable portfolio standards, international pressure to reduce carbon dioxide emissions, consumer preferences, and most of all, the lower price of natural gas, wind, and solar.

International coal use is expected to rise, at least in the short term. As nuclear power is scaled back across Europe in the wake of Japan's Fukushima accident, countries like Germany and Japan are building new coal-fired power plants.[6] As a reaction, the U.S. started to export coal to Europe to meet demand, increasing coal production in Appalachia. From 2000 to 2010, global coal demand grew an average of 4.7 percent per year, with demand in China and India leading the way.[7]

However, in 2019, Germany made the bold announcement that it would eliminate the use of coal for power generation by 2038.[8] Coal demand is projected to stall globally in the next few years due to shrinking demand from Europe and the U.S.[9] Likewise, other countries are expected to decrease coal use between 2014 and 2040.[10] Natural gas now costs about the same as coal in East Asia and South Korea, Taiwan and China are pivoting to Luquified Natural Gas (LNG).[11]

So what about clean coal and carbon capture? Even with strong policy support for clean-coal, the high costs of carbon capture and sequestration projects compared with cheaper alternatives are inhibiting its adoption. The Kemper County Plant, being built in Mississippi, switched to natural gas after exceeding its planned budget by $4 billion.[12]

Even the Department of Energy seems to have given up on clean-coal. It shut down its FutureGen carbon capture project that in

2004 it called "one of the boldest steps our nation has taken toward a pollution-free energy future." The Energy Department said costs that had risen from $1 billion to $1.8 billion were the chief reason for abandoning the project.[13] However, cheaper alternatives (such as natural gas and renewables) are also a likely culprit. This phenomenon is not restricted to the United States: a 2019 report by the Global CCS Institute found that the number of large-scale projects to capture and bury carbon had fallen from 75 to 43 worldwide, with only 18 in actual commercial operation.[14]

The economics of clean coal technology are very difficult. Efforts to capture and sequester the $CO_2$ requires not just additional hardware and capital expenses for the plant, but also energy to operate the capture equipment. And since the power plant is only one-third efficient to begin with, the additional power requirement for the carbon capture system reduces the amount of electricity generated and sold.

The decline of coal power generation in the U.S. is more than a passing trend. Even the EIA's estimates of coal still providing 13 percent of our nation's electricity generation in 2050 are likely overly generous. We anticipate coal will provide an even smaller fraction once utilities gain more comfort and familiarity with the alternatives.

During the first four and a half months of 2020, the U.S. was on track to produce more electricity from renewable energy than coal for the first time in well over a century. EIA projects that in 2020, coal will produce less electricity than both renewables and nuclear power.[15] Furthermore, planned closures of coal plants are accelerating, so it is not likely that coal will ever regain its dominant position in U.S. electrical generation.

## Natural Gas

In the first two decades of the 21st century, natural gas has been the primary winner in the electric generation industry. Gas should continue its success in the near-term, both in meeting new generation demands, and as supporting power for wind and solar as they expand. Modern, combined cycle, natural gas power plants are capable of operating at almost twice the energy efficiency of coal power plants. And since most of the potential pollutants in natural gas, such as sulfur dioxide, are removed prior to combustion, the emission streams from the plants are much cleaner than coal. This combination of higher efficiency, lower emissions, and now lower fuel price, have made natural gas power plants dominant in the power sector.

Gas is abundant, relatively clean at the point of combustion, and relatively inexpensive. Most experts agree that there is an enormous amount of gas in the U.S. alone. The reserves are estimated to be massive enough to last many decades, if not centuries.[16] And new hydraulic fracturing technology makes it relatively cheap to extract gas compared to getting to hard-to-reach conventional fossil fuel resources (such as offshore or remote fields). Because of these factors, U.S. gas production has grown significantly, starting in 2008. From 2007 to 2018, annual shale gas production in the U.S. grew from 1.3 trillion cubic feet to more than 17 trillion, and the EIA baseline forecast projects it to reach 33 trillion by 2050.[17] Because of this abundance and ease of production, natural gas prices have fallen steeply. But, following an increase in drilling and expansion into more difficult to reach reserves, recent estimates from the EIA have the natural gas price slowly increasing up to 2050.[18]

Though not expected to grow as fast as in the U.S., worldwide gas-fired electricity generation is also expected to grow by more than 50 percent from 2010 to 2040.[19] Whatever the demand for gas around the world, there are several significant reasons why gas production in other parts of the globe will lag behind that in the United States. For one, the geology of most of the rest of the world is simply not as well-known and mapped as it is in the U.S. Some early hopeful prospects for large deposits, for example in Poland, haven't panned out for a variety of technical, political, and cultural reasons. Much of the relatively new, highly specialized drilling equipment needed to get to the gas reserves is still tied up in the U.S. And, with their larger reserves of other fuels and other energy agendas, many foreign governments are unlikely to make natural gas production a high priority.

That said, China, Mozambique, and the Middle East all have huge gas reserves that could be tapped when the time comes. And under the sea, and beneath some permafrost, gas methane from hydrate sources could represent enormous reserves if the technology could be perfected to safely use it. At present, the Japanese are leading the search for just such a technology.[20]

## Nuclear

Nuclear has zero atmospheric emissions and is one of the lowest life cycle carbon sources of electricity available. So why aren't we producing most of our electricity using the abundant energy of the atom? Short answer: it costs too much and people are afraid of it. We're going to delve a bit deeper, but for the most part nuclear isn't the energy darling it once seemed because of its high capital costs and operational risks. In the U.S., nuclear power production has stopped growing. It's essentially flat, and on the brink of the first decline in its history.

After decades of no new construction, five new units started construction in the 2010s with the expectation of replacing several of the oldest and smallest units that have closed in recent years. The most recent setback came in South Carolina as the board of the utility Santee Cooper abandoned construction of two reactors at the Summer plant when construction costs soared to $25 billion, 75 percent higher than original estimates.[21] In addition, as much as a third of the current nuclear plants are at risk of closure because they're not economically competitive with natural gas, wind, or solar.[22]

According to the World Nuclear Association, almost half the nation's nuclear plants operate in deregulated markets where the electricity they produce competes with electricity produced by other means, and is sold to utilities and other suppliers at a daily auction.[23] Total generating costs have dipped in recent years, and the Nuclear Energy Institute reported that prices in 2018 were 25 percent below the peak costs of 2012.[24] Even so, new nuclear is still not competitive enough to win out against other, cheaper electricity sources.

As of January 2019, there were about 449 nuclear power reactors operating in 30 countries. About 55 reactors were under construction in 13 countries, with another 160 on order or in some stage of planning, although some of these plants have since been put on hold.[25]

The future of nuclear power globally is mixed, with some countries proceeding with many new plants and others closing existing plants and canceling plans for new ones. France, which has been getting almost 75 percent of its electricity from nuclear, says that they might close up to 17 reactors in order to meet a new law reducing their reliance on nuclear power to 50 percent.[26]

On the other hand, China and Russia are both building several new nuclear plants. In 2019, China had 45 units in operation and was building 15 more, with plans for another 30.[27] Russia has 38 plants in operation, 24 planned, and another 20 confirmed or planned for export construction.[28]

Future reactors are expected to be funded by governments or government-controlled companies, as is the case in China and Russia. With the exception of the United Kingdom, which has two units under construction and two more planned, very few new nuclear power plants are being built in competitive markets.[29]

Its impacts can be significant due to the volume of cement required for power plant construction, the electricity for fuel enrichment, and the waste handling challenges (which remain vexing). However, nuclear power generation produces no atmospheric emissions. In total, its full life-cycle carbon footprint is lower than that of coal, gas and hydropower, and roughly equal to wind and solar.[30] However, until the environmental benefits of nuclear and its ability to produce power around the clock are valued in the competitive market, its long-term future will remain uncertain.

## Wind

There's plenty of wind circling the globe to meet our power needs. A Stanford University study showed that wind power could meet world energy demand five times over.[31] Because of its ready availability, simplicity, and low cost, wind has been leading renewable energy growth, both in the U.S. and worldwide. U.S. wind capacity has risen from about 2.4 GW in 2000 to more than 97 GW in early 2019.[32] World wind capacity soared from just under 17 GW to more than 597 GW by the end of 2018.[33]

In the U.S., the cost of wind energy has plummeted in the last few decades, dropping from over 50 cents/kWh in 1980, to less than to 2 cents/kWh in 2017.[34] Prices for wind energy worldwide have dropped correspondingly, and are now competitive with coal and gas.

Transmission remains a primary problem for onshore wind because people typically don't live where it's windy. In the early 2000s, hundreds of turbines were built in remote west Texas, far from the cities that wanted the power they were generating. Transmission was the obstacle, as it still is in many parts of the world. Toward the end of 2013, much of Texas's transmission problem was addressed, as the $7 billion CREZ project was completed. CREZ, or Competitive Renewable Energy Zones, can today send 18,500 megawatts of wind power throughout the state, or three times as much wind power as any other state in the U.S.[35]

On December 27th, 2018, the Texas grid operator, ERCOT, was getting 54 percent of its energy from wind generation.[36] And for a brief time in 2018, the Great Plains electric grid, powering customers in 14 states, was meeting 60 percent of its requirements with wind energy.[37]

Similar long-haul transmission issues don't exist for offshore wind. Nearly 80 percent of the world's population (and load centers) reside within 200 miles of an ocean coastline making offshore wind a natural fit for colocating electricity generation with electrical load.

The leaders in offshore wind production are Great Britain, Germany, and Denmark.[38] Great Britain gets more power from offshore wind than almost all other countries combined, with more than 1,000 turbines. In 2017, the Dutch opened what was billed as one of the world's largest offshore wind farms in the North Sea, with 150 turbines, that could supply the energy needs of 1.5 million people.[39] In 2016, the U.S. opened its first offshore wind farm near Block Island, New Jersey, consisting of 5 turbines with a capacity of 30MW.[40]

In a signal of just how fast shifts to clean, safe, renewable energy can be made, Japan is building 140 massive wind towers 12 miles offshore from the ruin of the Fukushima nuclear plant. It's said this wind complex alone could produce over 1 GW of power by 2020.[41] And wind turbines are getting much larger. GE's newest offshore model, the Haliade, is a 12 MW behemoth installed off the coast of Rotterdam, with a 220m rotor diameter.[42]

# EVOLUTION OF WIND TURBINE HEIGHTS AND OUTPUT

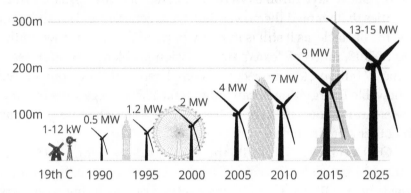

Source: Varous. Bloomberg New Energy Finance

**Figure 5.**

Many industry experts predict that the next great wind boom in the U.S. will be offshore along the coast of New England and the central Atlantic states. There are currently 12 active offshore wind leases being developed in the U.S. with a combined potential for 15 GW of generation. The future of U.S. offshore wind will depend heavily on the progress of these projects. But based on offshore wind's success in Europe, it is very likely that we will replicate that approach to great scale in the United States in the coming decades.

## Solar

Yes, wind could meet our energy demands five times over. But the sun dwarfs even that power. In fact, Sandia Labs estimated that the solar energy striking the earth's surface in less than two hours could easily meet the world's energy demands for an entire year.[43] Solar energy's unparalleled abundance, eminent renewability, and rapidly declining cost are driving exponential growth in capacity. This trend will undoubtedly continue, and solar will almost certainly become the leader in renewable energy growth in the coming years.

Solar PV panels have no moving parts, and aside from the carbon footprint associated with their manufacture and end-of-life management, produce zero emissions during the decades-long life of their operation. The sheer simplicity of a device that just sits in the sun and yet produces the electricity we all crave and depend upon is rapidly changing our relationship with energy. The sun is the primary source of most every form of energy we access, and solar PV panels provide us the means to harness it and transform it directly into electricity.

The solar PV panel has literally "democratized" power generation for the world. The simple, modular, scalable, and solid-state nature of solar PV has put electric power generation within reach of individuals with one or a few dozen panels, commercial and industrial consumers with hundreds to thousands of panels, and even traditional utilities with millions of panels. It is this simple universality of the technology that is driving its exponential growth, since almost any electricity consumer can take advantage of it.

Photovoltaic deployment might look like the adoption of a new consumer product such as a smartphone, not following the usual timeframes for standard large-scale power plants. Solar cells can be manufactured in a factory, shipped over conventional distribution systems like consumer electronics, and don't require the planning, permitting, construction, fuel acquisition, and operation and maintenance of a large utility power plant. And solar manufacturing continues to become more automated. First Solar has a new manufacturing facility that is now almost completely automated after originally requiring hundreds of employees.[44]

In 2008, the U.S. had 618 MW of solar capacity installed. Just 11 years later, solar capacity had expanded two orders of magnitude to 67,000 MW.[45] Much of solar power's rapid and accelerating growth is due to its low cost. From 2010 to 2017, utility-scale solar PV power fell from more than 20 cents per kilowatt-hour to under 3 cents.[46] From 2010 to 2016, the average per-watt cost of a solar PV system in the U.S. dropped by 15 percent per year.[47] Utility-scale installations fell to under $1 per watt in the first quarter of 2019.[48]

No other fuel used for power production—renewable or not—is predicted to see as much percentage growth as solar in the near-term. Its competitors are taking notice. Shell, a company once known almost entirely for oil, believes that solar will be the number one source of electrical power on earth by the end of this century.[49]

# 1980–2013 U.S. WIND COST AND POWER CAPACITY

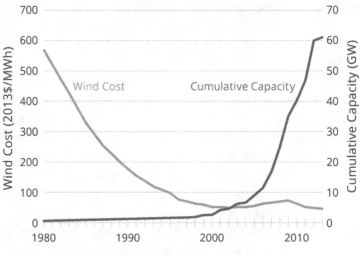

Source: U.S. Energy Information Administration

**Figure 6.**

# 1977–2015 GLOBAL SOLAR PANEL PRICES AND INSTALLED CAPACITY

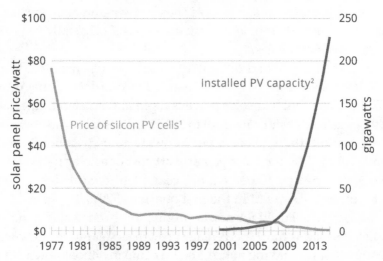

Source: [1] *Bloomberg, New Energy Finance & pv.energytrend.com;*
        [2] *IEA, 2015 Snapshot of Global Photovoltaic Markets*

**Figure 7.**

## *Hydroelectric power*

From the late 1800s through the midtwentieth century, the U.S. built tens of thousands of dams for flood control, irrigation, navigation, and hydroelectric power. The more recent trend in the U.S. is to tear down many of these larger dams, as environmental groups point to damage to land and habitat destruction. While there is still potential for a few new large, hydroelectric power dams in the U.S., there is significant environmental resistance. However, smaller dams and retrofitting of powerhouses at existing dams provide some opportunity for growth.

By contrast, the untapped global hydropower potential is substantial. Countries like Russia, China, Brazil, and India have significant room for hydroelectric power projects.[50] In Africa, the potential is astounding. Countries like Mozambique, Madagascar, Ethiopia, Democratic Republic of Congo, and Gabon see almost all of their potential hydroelectric power untapped. However, despite the allure of large, emissions-free sources of electricity, there are always tradeoffs; the large reservoirs from new hydroelectric power projects would likely cause environmental damage and the displacement of many people. Consequently, several South American countries have recently scaled back or canceled plans for large hydropower dams and are looking instead to wind, solar, and natural gas.

## *WHERE* WE GENERATE ELECTRICITY IS CHANGING: DECENTRALIZATION

In the early years of commercial electricity generation, Thomas Edison's direct current was the norm in the U.S. The primary electrical load in those days was lighting, which was a good match for direct current. But low-voltage direct current, like that produced at Edison's Pearl Street Station in Manhattan, couldn't be transmitted over long distances, and wasn't flexible enough to power the growing variety of electric devices and appliances that needed higher voltages.

In particular, direct current is difficult to step up and down in voltage, making it inconvenient for transmission and distribution. Consequently, power plants had to be located near the loads. During the first years of electrification, most buildings housed their own generators. Power production was distributed widely among industrial, commercial, and residential properties.

The outlook changed when Nikola Tesla, George Westinghouse, and a host of others contributed to make alternating current a viable option in place of direct current. Alternating current was easier to step up and down to higher and lower voltages. That capability enabled central power stations to be built outside of cities, as they could use transformers to elevate the voltage for efficient long-distance transmission and then drop the voltage as they distributed the electricity for use by consumers.

Thus, alternating current transformed the power industry into what it is today, with almost all consumer power produced at centralized plants far away and thereafter transmitted and distributed to homes, buildings, and factories. Simultaneously, economies of scale made those centralized plants the most efficient and cost-effective electric provider.

Today, technology is pushing the location of power production full circle. As we noted in the section on the future of buildings, distributed generation is becoming common again and many of the generation technologies are direct current devices. Distributed solar PV, diesel generators, gas microturbines, fuel cells, and other new power generation devices are becoming competitive with, and disruptive to, grid-priced electricity. Lower prices, the push to use renewable energy, and the need for resiliency are all driving this rise in distributed energy generation.

### Space Required

With the growth of distributed power, especially power produced by renewable sources, comes another issue related to "where." Because of their diffuse nature, wind and solar take up more space per kilowatt of capacity than fossil fuels. The reason is simple. While magnificently abundant, solar and wind are not fuels, but rather energy flows that we harness. Consequently, their energy density is significantly less than that of coal, gas, or nuclear, so more space is required to produce the same amount of electricity. Nuclear advocates point to this fact in support of more nuclear-produced electricity. Solar and wind, they posit, take up too much valuable land that could be used for crops and people or left in its pristine natural state.

However, much of the land for solar and wind can be used in a fashion benign to the environment and the economy because of its compatibility with other uses. Rooftop solar, for instance, is generally applied to space that is not being used for any other purpose. No one's planning to build a nuclear plant on a rooftop. Also, many of the world's solar farms are being built in vast deserts. If the fragile desert landscape is protected, then large cities could be powered from these solar farms. Wind farms are being built on mountain ridges or anchored offshore, or share ranch land with cattle in West Texas. So, while sometimes massive, these projects often take up little in the way of productive land area, or in the case of rooftops, any at all.

## Smart Grid

An important part of "where" power is generated is getting that power to the places it needs to go. Most all of us are familiar with "the grid" and its traditional role in moving power from centralized power production sources to our homes, shopping centers, and places of work. Today, most see a "smart grid" evolving from the old-fashioned one.

Smart grid has become a general term referring to different aspects of a changing electrical system, on both the utility and customer sides of the meter. Advanced metering devices, distributed generation sources like solar and fuel cells, energy storage and energy efficiency devices, electric vehicles, billing systems, demand response devices, and tariffs: they're all part of our quickly evolving smart grid.

Some of the upgrades to the transmission grid are necessitated by the impressive expansion of renewable-based energy production in far-flung locations that will require investment in transmission capacity. Likewise, distribution system upgrades are necessary, as rapidly expanding distributed energy is available to feed power back into the grid. Yesterday's transformers and circuits were chiefly designed to send electricity in just one direction. They are bulky, often cooled with oil, and not designed to deal with rapidly changing loads.

There are also opportunities for conservation by giving customers better and quicker feedback on their consumption patterns. Smart meters are usually the first step in these upgrades. These meters allow the utility and the customer to receive shorter measurement intervals than a typical monthly utility reading. Smarter meters vary in capability, with the more advanced referred to as AMI, or advanced metering infrastructure. These kinds of meters are a necessary platform for many energy saving strategies, such as time-of-use rates or dynamic pricing.

The evolving smart grid will have to handle electricity that flows to and from multiple distributed generation locations. Solid-state transformers are a likely necessary technology to make this advanced smart grid work. Solid-state transformers can both scale down voltage to loads such as homes and scale up voltage from sources such as solar panels and electric vehicle batteries. Furthermore, solid-state transformers can be about a tenth the weight and a quarter the volume of traditional transformers.[51]

Utilities and outside investors are reacting swiftly to meet the needed changes. The global smart grid market is expected to reach almost $51 billion by 2022, and the U.S. is expected to hold the largest market share.[52]

## *Microgrids*

Microgrids are small electrical grids that can "island" or disconnect from the main electrical grid and operate independently. Microgrids are often seen as more sustainable than the traditional grid, since many use local, renewable energy resources to provide their generation. Such isn't always the case, however, and many of the largest microgrids currently operating are fueled by natural gas.

Microgrids have received a good deal of attention due to their perceived resiliency. In the wake of Superstorm Sandy, for instance, Princeton University did not lose electrical power because of its microgrid. The chief advantages of a microgrid are its resiliency and its flexibility to be powered renewably or by fossil fuels. But they are not invulnerable. Microgrids can go down themselves when hit directly by weather or disaster, as the buildings in such a grid are still connected by conductors, either in overhead wires or underground vaults. Thus, storms with high winds or flooding could disable a microgrid even if it's been "islanded" from the main grid.

## *WHEN* ELECTRICITY IS GENERATED AND CONSUMED IS CHANGING: STORAGE

The electric utility industry has always been a "just in time" business. For nearly its entire history, electricity generation had to be nearly instantaneously balanced with electricity load to maintain the grid's voltage and frequency and prevent power outages. Energy storage is changing that.

Traditionally, the only utility-scale energy storage has been pumped hydroelectric storage, where a utility pumps water uphill into a reservoir during off-peak hours and releases it through generators during peak hours. Worldwide there are presently more than 300 pumped hydro projects capable of generating more than 153 GW of power. These projects account for almost all of the current utility-scale storage.

These pumped hydro facilities are essentially hydro power plants and require favorable terrain features and access to large quantities of water. Even though they don't require the damming of waterways, their reservoirs flood large areas and can have significant impact on local ecosystems. This is why many environmental groups resist them. But they are simple and clean at the point of generation and do not use exotic materials typical of many modern batteries.

Pumped hydro is a form of gravitational energy storage. But raising water to a higher elevation is just one of many ways to store energy via gravitational potential. Several firms are experimenting with moving heavy concrete masses to higher elevation. One approach uses railroad tracks leading up the side of a gradual hill. Excess or off-peak electricity is used to run electric locomotives up the gentle grade towing massive concrete or rock-laden rail cars. When the stored energy is needed, the locomotive uses regenerative braking to slowly run back down the hill, converting the gravitational energy back into electricity.

Just as simple, but a bit more novel, another young company is experimenting with traditional construction cranes that stack massive concrete blocks in a concentric tower around each crane. Again, excess or off-peak electricity is used to power the electric cranes as they stack the blocks into higher and higher towers. When the stored energy is needed, the cranes operate in a reverse regenerative mode deconstructing the towers and generating electricity as the blocks are lowered back to the ground. While they have low energy density, the simplicity and modularity of both approaches means they could potentially be scaled up. This storage technology may particularly benefit places without water for pumped hydro.

Another large-scale storage option is Compressed Air Energy Storage (CAES). When off-peak power is used to drive an air compressor to force compressed air into an airtight underground cavern, like a natural salt dome, that air can be released on demand at peak-time to help power a gas turbine. While CAES works well, it requires the proper geology for underground air storage, which isn't universally available. As of 2019, there were only two CAES facilities in operation, one in Germany and one in the U.S.

Utility-scale battery storage has been growing rapidly. Utility-scale batteries are being tested in many forms: sodium-sulfur, lithium-ion, sodium-ion, and flow and air batteries. Other non-battery storage solutions, like liquid air, molten salt, and flywheels, are also under development. Utilizing the storage capacity of the many electric vehicles connecting to the grid is also proposed as a form of distributed energy storage that can reverse the charging process in what is known as V2G (vehicle-to-grid).

The federal and state governments have started pushing energy storage technology. The Federal Electricity Regulatory Commission (FERC) passed FERC Order 841, which "requires wholesale markets to create a 'participation' model for electric storage resources."[53]

The energy storage industry was given a substantial boost by the state of California, which mandated that 1.325 GW of storage be deployed by 2020. That number represents a 50 percent increase in worldwide nonhydro energy storage capacity. Several states including Utah, Oregon, New York, and Massachusetts have mandated some level of energy storage on the grid.[54]

Several firms are piloting battery storage systems up to 10 MW to be used on-site at wind farms or within the distribution system itself. At present, there are more than 150 companies delving into this increasingly competitive side of the energy equation. In late 2017, Tesla paired a 100 MW battery storage facility with a wind farm in Australia, marking an important threshold for size and speed of construction. The battery cost 66 million dollars and reportedly made 17 million dollars in the first 6 months of operation. The Australian Energy Market Operator said it was more rapid and accurate than a conventional steam turbine.[55] And Florida Power and Light has announced plans to build a 409 MW battery to store power from their solar power plant in Manatee County. It will be four times larger than any other battery system in operation.[56]

Molten salt is becoming a preferred storage medium for concentrated solar power plants. By storing a large portion of the energy they capture during the day in molten salt, these solar thermal power plants can then extract the heat at night to continue powering their thermal electric generation plant and provide "solar power" long after the sun has set. Thermal salt-based storage has the potential to be competitive with lithium-ion batteries and other grid scale storage.

Arizona's Solana project already gathers solar energy and pumps it into huge tanks of molten salt. When the sun goes down, they can pull heat out of the molten salt and into steam generators.[57] Alphabet, the parent company of Google, has a project using salt and antifreeze, and Siemens is also developing salt-based storage for its solar-thermal plants.

Many utilities use chilled water for thermal storage, cooling water during off-peak hours and using it to reduce peak loads on air conditioners by pumping it through buildings. Austin Energy has a large chiller plant located in the downtown core that services a convention center and many office buildings. It is cheaper for both the utility and the customers than generating or purchasing electricity during peak load hours.

Other distributed storage approaches include Ice Energy's Ice Bear units. Located on rooftops mostly in California for commercial applications, their Ice Bear thermal storage units use electricity at night to create ice, which is then used to assist air conditioning systems during the day.[58] In 2017, Ice Energy ventured into the residential market with their new unit, the Ice Cub, which replaces a conventional AC unit and provides similar efficiency gains to that of the commercial scale equipment.[59]

The bottom line: there are many approaches to energy storage, many novel technologies are in development, and we can expect a wide variety of energy storage deployed sooner rather than later. This build-out is important for the expansion of renewable energy resources such as wind and solar. The variability of these resources can be addressed with storage, rather than relying on "firming" power by fossil fuel power plants.

## *WHY* WE GENERATE ELECTRICITY IS CHANGING: WORKLOADS

The electrical loads served by the power industry are undergoing significant, and somewhat paradoxical, changes. On the one hand, the historical growth in electrical load in the U.S. peaked in 2007 and has essentially flattened out since, even declining in some areas. On the other hand, global electrical demand continues to increase.

Many different types of work in our economy are being electrified. Electricity is efficient to transport; can be produced by a wide range of sources, including fossil fuels, nuclear materials, and renewables; can be stored; and can be delivered at a wide range of power levels to finely match our diverse workloads.

The growing electrification of everything has been a consistent and continuing trend in electricity demand worldwide, increasing from just less than 15,000 billion kWh (15,000 TWh) in 2000, to almost 25,000 billion kWh (25,000 TWh) in 2018.[60] However, in the U.S., electricity generation peaked at more than 4,100 TWh in 2007.[61]

Since 2007, substantial energy efficiency improvements in appliances and buildings in the U.S. have worked to level and even reduce electricity demand in spite of a growing economy, expanding population, and more electrical devices. And many regions of the country are planning for even lower consumption in the next few years. But there are significant new electrical loads in the future that could change this picture.

According to a study by the Brattle Group, electric demand for utilities in the United States could double from 2015 to 2050. These new loads include charging for electric vehicles, electrifying air and water heating in buildings, industrial electrification, distributed manufacturing, indoor agriculture, and other uses.

## Electric Vehicles

Electric vehicle charging in particular could become a major share of the residential load as more EVs are sold, and as energy efficiency reduces traditional heating and cooling loads. The load created when an electric vehicle plugs in is the largest new load to appear in homes in a generation.

The Brattle Group concluded that total electricity demand could increase as much as 56 percent, over the 2015 load, if battery electric vehicles replaced the majority of America's light-duty vehicles.[62] Fortunately for the grid and the home owner, our vehicles tend to sit still overnight, and thus, charging them overnight is a benign electrical load as we and most of the large appliances in the home are asleep. That means electricity could expand significantly without severely impacting peak demand.

Between 2011 and 2018, the amount of electricity consumed by electric vehicles globally increased more than 260-fold. The IEA projects that electric vehicle electricity use could further increase five-fold by 2030.[63]

Complete electrification of our roughly 250-million-strong, light-duty vehicle fleet in the U.S. could add 700 TWh of new electrical demand.[64] As daunting as that sounds, it would only be a 17 percent increase over our current annual generation. In fact, if people are disciplined and charge only at night, that gross amount of new generation could be met without building a single new power plant! But it would mean running our existing fleet of coal and natural gas power plants at near full capacity. In other words, our theoretical fleet of electric vehicles would essentially be coal and natural gas burners. Thankfully the power fleet gets cleaner with time.

We won't all switch to electric vehicles tomorrow. And even when electric vehicles make up the majority of new sales, it will take decades for the full fleet of cars and trucks to turn over into a fully electric fleet. There is more than enough time to build sufficient new wind, solar, and nuclear generation to make it a truly "emissions free" fleet of electric automobiles.

## Data Centers

As one might expect, computers represent a growing share of world energy consumption, even as they grow more efficient. Data centers, with their stacks of computers, are now major electrical loads. According to a study by Datacenter Dynamics, global power demand by data centers grew from 12 GW in 2007 to 40 GW in 2013.[65] The internet may decrease some forms of energy consumption, but is itself a tremendous new electrical load. Though the Semiconductor Industries Association anticipated that unrestrained growth in electrical demand by computers will consume more energy than the world produces by 2040, efficiency measures have severely limited that growth.[66]

One computer demand that produced headlines is cryptocurrency, using the underlying blockchain concept. We will not attempt to explain how blockchain and cryptocurrencies work, but focus instead on the resulting electrical demand. The "proof-of-work" concept requires the solution of increasingly complex mathematical problems, and successful "mining" of crypto coins comes down to more computational power using specialized computer chips. The electrical loads are almost unbelievable.

In December of 2017, the estimated annual electricity consumption for mining Bitcoin, just one of several cryptocurrencies, was equivalent to the annual electricity consumption of Serbia.[67] Ethereum is touted as a much more energy efficient coin, but in 2017 Ethereum used the equivalent of Sri Lanka's consumption.[68]

However, these estimates are based on one data source— Digiconomist—and some have questioned the consumption as being too high, as it is based on revenue and operating cost estimates of the miners, rather than actual electricity metering. And cryptomining should benefit from the same advances in computer efficiency that has disproven the initial energy claims made about data centers. The specialized cryptomining computer chips are becoming increasingly efficient, blockchain systems will shift to simpler algorithms, and cryptomining should enjoy the same efficiencies of scale as data centers.

## Indoor Agriculture

In the building chapter, we talked about the rise of urban farms. With their use of grow lights and pumps for circulating water, they represent a significant, new electrical load. In some states where recreational marijuana use has been legalized, the latest significant, new electrical load has been marijuana-growing operations. A typical "grow closet," accommodating 4 plants at a time, consumes as much electricity as 29 new refrigerators. A "grow house" can contain 10 to 100 modules.[69] Denver saw the cannabis industry account for about 4 percent of the city's electric demand in 2017.[70]

Indoor agriculture utilizing old shipping containers has also taken off in recent years. Companies like Freight Farms and Growtainer convert shipping containers into vertical farming set-ups. However, while these container farms have many benefits, a single container from Freight Farms can use about 80kWh of energy per day.[71] That is more than twice the average consumption of a home. As of 2017, there were several hundred container farms in the U.S.[72]

It is pretty clear that lowering our electricity consumption through improvements in energy efficiency and retirement of old loads could be offset by growth in new loads, many of which have yet to be considered.

## *WHO* PROVIDES US ELECTRICITY IS CHANGING: BUSINESS MODELS

So, who will be responsible for electricity generation? At first thought, even with all the change outlined in this section, it would seem that the utility company is still generating electricity and the consumer is using and paying for it. But the traditional electric utility business model is facing its biggest challenge since its creation over a century ago; a problem commonly referred to as the "utility death spiral."

The simple version is that customers are starting to generate their own electricity. As one customer reduces their purchase of grid-supplied electricity through rooftop solar generation, the utility has to raise rates on remaining customers to recover fixed costs for the grid. That encourages others to put solar on their roofs and purchase less, and the utility must raise rates again, thus the "death spiral."

Although this is often discussed in terms of "grid defection," the customers usually do not actually disconnect from the grid, but do significantly reduce their consumption from the electric utility, while still using grid-supplied electricity when necessary. These customers with the rooftop solar are the new prosumers.

Futurist and writer Alvin Toffler coined the term "prosumer," referring to members of an underground economy that performed work as well as consuming what they produced. Prosumer is a useful word for building owners who have distributed generation on the roofs, and who are now electricity producers as well as consumers. There's no longer simply a utility and a customer, but instead there's a complex relationship between two or more prosumers.

### New Business Models

Many companies and regulators are responding with new proposed business models and regulatory mechanisms. The New York Reforming the Energy Vision (REV), the United Kingdom RIIO (Revenue = Innovation + Investment + Outcomes), regulatory proceedings in California and many other states, nonprofit organizations like Rocky Mountain Institute e-lab, and academic efforts like the Lawrence Berkeley National Lab series are all examples of this ongoing effort.

In general, utilities are envisioned as moving to one of two new business models: the energy services utility, where the utility concentrates on providing services to the customer rather than electricity, or the smart integrator model, where the utility integrates distributed energy resources that are customer owned. Some companies like ENGIE, a multinational energy company based in Paris, France, where Michael works, are seeking to do both.

The common wisdom is that utilities will survive by cutting costs, creating a more valuable platform of services and finding new sources of revenue such as electrification of the transportation sector. In parallel, they might need to develop entirely new ways of getting paid.

### Platform Model

Regulators are encouraged to reward utilities for incentivizing distributed energy resources and energy efficiency, and cutting costs through using customer-owned resources. The most popular business model in the business and regulatory community has been to encourage utilities to move to the "platform" model that has proven successful with the new information-based businesses such as Apple,

Microsoft, and Google, and now applied to other services such as Uber and Airbnb.

The NY REV effort has specifically promoted the platform model for New York utilities, requiring them to file distribution service implementation plans using Distributed Energy Resources (DER) to cut costs, providing earnings adjustment mechanisms for efficiency and DER deployment, and envisioning that new products and services will eventually provide the majority of utility income.

The new models also require the emergence of electric distribution markets where customers can buy and sell electricity from customer-connected devices or in peer-to-peer markets as well as the grid, with the utility acting as a market platform. Finally, some models explore the possibility of replacing private distribution system operators with nonprofit independent system operators.

There are certainly reasons to believe that these new models can address utility problems, as least in the short-term. Some utility investments have been offset by efficiency and distributed resources, and there is certainly significant revenue potential from expanded electricity sales to support electric vehicles. One study showed the potential as several times the revenue loss expected from nonutility-owned solar PV generation.

However, there are several reasons that the platform model may not work for electric distribution utilities in the long run, especially in a scenario of low overall load growth and high penetration of distributed energy resources, according to a study by the LBJ School of Public Affairs.[73] These reasons include difficulty in finding new products and services that produce significant revenue, and the inability to charge a premium for connection to the grid.

Distribution markets, where the utility facilitates transactions among customers and takes a transaction fee, also has limited potential. It is assumed that distributed solar PV will be the prime energy source for these markets; however, solar-dependent distribution markets are going to be seasonal and locational in nature. There is simply not going to be a robust solar-powered distribution market during a gray, snow-covered New England winter.

Finally, unlike telecommunications systems, the value of distributed energy resources to the grid decreases with distance and market penetration. The best use and highest value for electricity generated from a distributed generation system, solar or otherwise, is on that site: the building itself. There is no transmission or distribution system involved. The second-best use for the electricity is to neighboring buildings, which automatically occurs when power is pushed back onto the grid. The third best use is for balancing the local grid, and then aggregating power for use balancing or meeting power needs further from the building.

The further you go from the building site holding the generation, the less value it has. The power cannot be used at all for sites that are not connected. Solar on my home in Austin simply cannot be delivered and used for a residence in Seattle unless I charge a battery and transport it to Seattle.

Similar constraints apply to market penetration. The value of my smartphone increases the more smartphones that are connected to the system. But this is not true for distributed power systems. The value of the electrical power from my solar rooftop system does not increase as more and more people put solar on their roofs. In fact, it decreases as far as usefulness to the grid. The value to my own home increases or decreases in relation to the price of electricity from the grid, not market penetration of solar.

All this means that the new business models may work for utilities in a low-penetration scenario such as today, but provide no guidance in a future where we have a massive deployment of distributed energy resources. A high concentration of solar and efficiency would be beneficial for customers, the environment, and society, but not necessarily profitable for electric utilities. In such a scenario, the best option for distribution electric utilities may be a fully regulated cost-of-service model that recovers fixed costs and allocates a fair rate of return to an independent system operator.

### Other Models

A new business model developed in California may be an example of a model that can survive in some future situations. Community Choice Aggregation (CCA) is where the utility continues to own and operate the poles and wires of the distribution system, but elected officials for a local government make the decisions on buying and selling power for their jurisdiction. The utility continues to bill the customers, but breaks out transmission expenses vs. generation paid to the CCA. Controversy revolves around the transparency of these charges and the need to make utilities whole for long-term contracts they entered to meet state-mandated renewable goals.[74]

## CHAPTER 7 SUMMARY

1. Our power generation is decarbonizing, as we move from fossil fuels to renewable energy.

2. Our power generation is decentralizing, moving from large central power plants to smaller distributed generation.

3. Energy storage is changing when we generate and consume electricity.

4. Computers, electric vehicles, and indoor agriculture represent significant new power demands.

5. Electric utilities need to change their business model to accommodate distributed ownership of power generation.

Planning goals should
be formulated around this objective:
What Reduces the most Greenhouse
Gases in the Shortest Time at the Least Cost.

Sam and Susan were concerned about the storm, but decided to go ahead with the party anyway.

"We've been planning this for weeks," Sam said. "Besides, we live in Houston. Storms come with the territory, like taco trucks and the Astros."

Sure enough, the party had gone on, and sure enough, the storm had taken down major transmission lines from Houston's wind and solar farms outside the city. The renewable energy power plants had plenty of energy storage, but no way to get the power to Sam and Susan's house.

Inside their house, the lights never even blinked as the neighborhood microgrid—energy stored from the community solar panels, local fuels cells, and a small, modular nuclear reactor nearby—kicked in. The larger citywide grid would be down a day or two, according to the radio, but the mostly underground microgrid held up and the party went on. The music playlist picked up where it left off and the frozen margarita machine hummed back to life.

The only downside was the ribs Sam had looked forward to barbecuing outdoors; it looked like frozen pizza would be the storm entrée.

# CLEAN ENERGY SOLUTIONS

Generating power in a sustainable fashion means producing it more efficiently and cleanly. That means meeting the world's energy needs while preventing pollution, including greenhouse gases. The goal might sound daunting, but there is good news. The natural supply of carbon-free energy is abundant. We could readily meet all the world's energy demand using either solar or wind power. Add in other carbon-free sources such as tidal, hydro, geothermal, and nuclear energy, and the options for a sustainable future expand. We have the technology to harness these forms of energy today. We've been harnessing solar, wind, and other renewable energy sources for decades, and can do so efficiently.

The even better news is that we're already trending in the right direction. The displacement of coal by natural gas has caused the most substantial decarbonization in U.S. history. An ongoing wave of coal generator retirements will continue that trend, taking tens of gigawatts of additional carbon-intensive capacity offline.[1]

According to global statistics, the total renewable capacity added in 2018 represented greater than half of all net additions—more than coal and gas capacity additions combined. And for the first time, there was more solar capacity added than any other generating technology.[2]

The bad news? According to most experts, we're still not doing enough. We're falling short of the goals most scientists feel are necessary to avoid serious climate change impact. The statistics for total energy supply and electrical generation worldwide don't show carbon-free, energy-based production and efficiency increasing anywhere near the rate needed to avoid widespread, climate-based problems.

From 1990 to 2015, world total primary energy supply grew at an annual rate of 1.8 percent. During that same period, renewable energy grew only slightly faster, at an annual rate of just 2 percent.[3] By 2018, wind, solar, and geothermal provided only 9.3 percent of the world's electricity production.[4]

Years ago, the IEA developed measures to track the target goal of limiting global temperature increase to only 2°C above preindustrial levels, commonly known as the 2D goals. Failure to meet these goals could result in serious climate disruption. These measurements of power supply and use include renewables, nuclear, gas, fossil fuels with carbon capture and sequestration, building efficiency, appliances, and more. As of 2017, only four were on track: electric vehicles, energy storage, solar PV, and offshore wind. Efficiency measures for buildings, albeit respectable, are not advancing quickly enough to constrain energy consumption growth in the residential and commercial sectors.[5, 6]

Even worse, the percentage of total energy supplied by fossil fuels globally is about at the same level it was 25 years ago.[7] All the new solar and wind power has just been added to a larger total consumption, and much of the emission benefit of the new renewables has been offset by the closing of nuclear plants.

The world is lumbering toward severe climate change. The causes of inaction include entrenched economic interests and the huge size of the power system, which slows the rate of infrastructure change. Scientists agree that if we are to reach an overall solution, we need to pursue many disparate clean and renewable technologies simultaneously.

## WEDGES

One such approach that involves a diverse set of solutions is called the Princeton Wedge approach. In a 2004 *Science* journal article,

Princeton professors Rob Socolow and Stephen Pacala, co-chairs of the Carbon Mitigation Initiative and the Princeton Environmental Institute, introduced the wedge approach to address the climate change problem.[8] This approach is a blend of end-user efficiency and conservation measures, alternative energy sources, power generation changes, and agricultural and forestry plans.

The wedge is a description of what a carbon reduction approach looks like when plotting carbon emissions vs. time. In the business-as-usual case, carbon emissions continue and grow with time. However, each reduction from business-as-usual reduces the growth in emissions, creating a reduction-wedge between the business-as-usual scenario and the reduced-emissions scenario.

The wedge approach is intended to address the difficult task of determining which solutions, or mix of solutions, can reduce greenhouse gas emissions at a scale that would help the world best avert the most potentially devastating impacts of climate change. This approach identifies 15 stabilization wedges that could potentially reduce greenhouse gas emissions by one gigaton each within 50 years.[9] The IEA tracking system mentioned earlier is also a wedge approach.

## STABILIZATION WEDGE THEORY

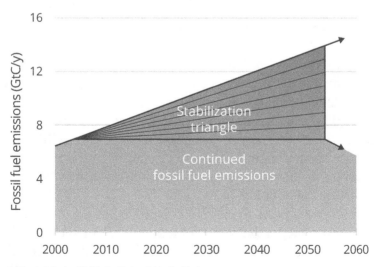

Source: S. Pacala, R. Socolow, "Stabilization Wedges: Solving the Climate Problem for the Next 50 Years with Current Technologies," Science (August 2004)

**Figure 8.**

Another very aggressive approach, from Stanford University, is called the WWS solution, where all our energy needs are met from three sources: wind, water, and solar. According to Professors Mark Jacobson and Mark Delucchi, "We don't need nuclear power, coal, or biofuels. We can get 100 percent of our energy from wind, water, and solar (WWS) power. And we can do it today—efficiently, reliably, safely, sustainably, and economically."[10] Jacobson's plan would involve building a massive number of new systems to generate 100 percent of our electrical power:

- 3.8 million wind turbines, 5 MW each, supplying 50 percent of the projected total global power demand;

- 49,000 solar thermal power plants, supplying 20 percent;

- 49,000 solar PV power plants supplying 14 percent; and

- 1.7 billion rooftop PV systems, 3 KW each, supplying 6 percent.

  The remaining 10 percent would come from a combination of geothermal and hydroelectric power plants, ocean-wave devices, and tidal turbines.

Jacobson points out that we would also need to build a massive new transmission infrastructure while simultaneously reducing demand for electricity.[11] He published a follow-up study that divided the world into 20 regions and used energy storage solutions unique to each region. The report says the renewable energy is 75 percent less costly than a traditional generation plan. Much of the cost savings are achieved through avoiding health issues that would occur from breathing polluted air.[12]

There have been significant objections raised to the Delucchi and Jacobson model, pointing out modeling errors and implausible assumptions.[13] In fact, the model does seem to break down for the last 10 percent of power needs, particularly the hydropower calculations and assumptions regarding underground thermal energy storage.

But whether it is technically feasible to achieve 100 percent of our energy needs from renewable energy resources is almost beside the point. The claim of most of these approaches (including Jacobson's) is that not only can we achieve a 100 percent renewable energy-based economy, but that it can be done within a few decades, will not require major changes in our lifestyle, and will actually be cheaper to achieve than continuing with a fossil-fuel based economy.

Unfortunately, we do not think this is the case. Although we believe it is both possible and imperative to achieve a 100 percent (or near that) renewable energy economy, we do not think it is going to be easy, achieved within a few years, or cheaper (at least in the near term) than fossil fuels.

The practical problems with implementing these "100 percent" goals and plans are discussed in *Our Renewable Future* by Richard Heinberg and David Fridley of the Post Carbon Institute. They point out six reasons that the transition will be difficult: intermittency of renewables, the liquid fuel problem (for transportation), uses for fossil fuels other than energy, the area density of energy collection activities, the location of renewable resources, and the quantity of energy provided. And they conclude that we, indeed, will need to make changes in our lifestyle, including shifting energy use to times of abundant supply, and perhaps less mobility. And even then, we will still find it difficult to find substitutes for fossil fuels in high-temperature manufacturing and as chemical feedstocks.[14]

It is laudable for cities, states, and nations to posit 100 percent renewable goals, but the reality of achieving 100 percent quickly or affordably is almost impossible. The reason is simple: as with many things, achieving 80 or 90 percent of anything often requires only half the effort, time, money, etc., of achieving the last 10 or 20 percent. In fact, in cases such as 100 percent clean and renewable energy, achieving the last few percent become much more expensive if we expect to maintain the same reliability.

But even with substantial obstacles, there is a climate imperative to get as close to a 100 percent carbon-free economy as quickly as possible. We propose a different goal, that we will call "80 percent as soon as possible." We believe this approach will avoid some of the problems associated with establishing 100 percent goals by a certain date. That buys us time to figure out more elegant solutions for the final 20 percent, which might be combinations of direct air capture, offsets, or other approaches.

## 80 PERCENT ASAP

First, this goal will avoid the arguments over the technical feasibility of getting to a 100 percent renewable economy. There is very little technical argument that we could not achieve an 80 percent goal.

Second, we need to achieve the 80 percent goal as quickly as possible to slow down the worsening effects and buy ourselves some time. We should not focus on any particular date in the future as "the end of the world" if we do not achieve it. The reality is that the climate is going to continue to get worse over the coming decades no matter what we do at this point. It is not the case that if we transition to a 100 percent renewable economy in the next 11 years then everything will be fine, nor is it the case that if it takes 13 years then the world will end. What we are really facing now is limiting the extreme of climate change and reducing the time it takes to return the atmospheric concentration of greenhouse gases to preindustrial levels. Remember our admonition in the introduction to this book that projections of what will happen by certain dates will almost certainly be wrong.

Third, this goal will shift the focus on cost to prioritizing what is achievable with current resources. We think planning goals should be formulated around this objective: *what reduces the most greenhouse gases in the shortest time at the least cost.* As a practical matter, the programs we need to implement over the next several years are the same whether the goal is 100 percent by a certain date, or 80 percent ASAP. Nobody is going to start with the actions needed to get the last and most expensive 20 percent reductions.

Fourth, this goal focuses on mass deployment of existing technology, rather than research and development. We can transition to an 80 percent carbon-free economy with the wind, solar, energy efficiency, and other technologies that we have today. We do not need innovation to solve the climate crises, although innovation will make our technology more effective and cheaper. But actions taken in the next couple of years will have a greater impact on reducing climate damage than actions taken ten years from now, even if the technology is cheaper and better then.

## CLEAN ENERGY PRINCIPLES

After looking at all the problems we face in developing a supply of clean, affordable, and reliable energy, it would be easy to start proclaiming the virtues of some energy sources and declare that the solution to our problem is "X." Much of the political discourse around energy uses this approach, as people line up in support of or in opposition to specific fuels and technologies they happen to love or hate. By now, we assume that the reader understands that we believe the problems to be much more complicated than can be addressed by simplistic one-off solutions or the promotion of any particular fuel or technology.

While there are a multitude of energy sources and technologies that can be considered clean, there are four general principles we recommend using for clean energy planning:

1. Meet workloads without energy conversion wherever possible.

2. Use energy sources and technologies without pollutants.

3. If polluting energy resources must be used, minimize their use.

4. Capture and reuse or store all life-cycle emissions.

A prime example of how to apply the first principle is with passive solar architecture in our homes, buildings, and factories. When the design of a house allows you to use daylight for reading rather than turning on light bulbs, then you are using a natural energy flow (sunlight) to meet the work required (illumination for reading) without an energy conversion. The same can be said of hanging clothes on a line to dry, sunlight heating up a stone floor in winter that provides heat later in the day, placing a six-pack of beer in a cold mountain stream to chill it, opening a window to ventilate buildings, etc. Because every energy conversion leads to losses, avoiding those conversions in the first place is paramount.

There are obvious energy sources such as solar, wind, tidal power, and nuclear which do not emit pollution at the point of conversion. Wherever possible, these emissions-free sources should be used. However, the fabrication of the devices that convert those pollution-free resources into electricity involve metalworking and chemicals that invariably produce pollutants in the process. Also, the specific location of these technologies must be judged on site-specific criteria for environmental impact.

Finally, if one could capture and store all emissions produced during the manufacture and operation of an energy source—including the processing of the fuel, the manufacturing of the energy conversion technology, and the operation of the power plant—then the entire operation would be an example of clean energy. This is very difficult in practice.

For example, the claims for "clean coal" focus only on the capture and sequestration of air emissions from the smokestack of the plant, not the mining, processing, and transportation of the coal or the manufacturing of the power plant components. Nuclear energy is declared to be clean because there are no emissions from the "stack," but ignores the mining and processing of the uranium, the steel and concrete and other components of the plant, and the handling of the radioactive spent fuel. Solar and wind proponents claim them to be pollution-free, but ignore the steel, concrete, and aluminum going into foundations, towers, and frames, and the toxic chemicals used in the manufacture of the solar cells.

In fact, we do not know of any technology for the generation of electricity or production of a transportation fuel that is pollution-free. While it is theoretically possible to capture and store all the emissions of these processes, it would be extremely difficult to do so in practice.

Here are some additional suggestions for achieving the most reductions at the least cost in the shortest time:

Minimize the infrastructure change. We have built a multitrillion-dollar global energy conversion and distribution infrastructure, and any time that infrastructure can be used for renewable energy without modification or minimal modification, the result is going to be a cleaner and cheaper method for meeting work needs. "Drop-in" fuels that can use existing fuel processing and distribution infrastructure will have a cost and time advantage over energy and technology combinations that require new infrastructure development.

An example would be electric vehicles in comparison to other alternatives for transportation fuels to replace petroleum. Whereas some biofuels and hydrogen require a whole new fuel production and distribution infrastructure, the infrastructure is largely in place for electric vehicles. Power plants have excess capacity in the evening hours for charging, and the electrical distribution system already exists. Likewise, if a biofuel was developed that was a "drop-in" into the current fossil-fuel pipeline distribution system, it would have an advantage over a fuel that could not use the current infrastructure.

Additionally, we should use energy flows when they are available. This approach is in contrast with our current paradigm, which is to use energy whenever we want and then to ramp the energy infrastructure up and down to follow our demand. But what if we switched our demand to follow the supply rather than changing the supply to meet our demand? The rise of flexible industrial processes (such as water treatment, data centers, electro-fuels manufacturing, and so forth) could be modulated to match when clean energy is available. Sunlight, wind, flowing water, tides, and a few other forms of energy flows can be converted to useful energy as they occur without the need for elaborate fuel discovery and processing. Doing so also minimizes requirements for energy storage. The most efficient use of the energy would be to meet a current workload with no or minimal conversion, and that means avoiding storage if possible.

We should also try to match our workloads with fuels that have the appropriate energy density. Rocket fuel is an example where the weight and energy density of the fuel is appropriate to the workload, lifting a rocket into space. Another example would be using a fossil fuel to produce the heat necessary to make cement. On the other hand, illumination does not require the construction of a large power plant, and can easily be met with on-site solar and wind technology.

In particular, assessing workloads in terms of high- and low-density power needs and matching them to appropriate energy sources would lead to more efficient energy conversions. The power that can be extracted from renewable fuels directs them more toward relatively low power residential uses, for example, than most industrial applications. And liquid fuels, such as biofuels, are well matched with the workloads of certain segments of the transportation sector that cannot easily be electrified.

We should also convert our work processes to electricity whenever possible. Electricity is a low entropy form of energy. Heat is a high-entropy form of energy. Electrification of work processes will usually be more efficient and less polluting than the nonelectrical process it replaces. For example, steel mills that have switched from burning coal to electric arc heating have improved their competitiveness while reducing energy requirements.

There are also nonobvious pathways to saving energy. For example, providing clean water is very energy intensive end-to-end from the sourcing, pumping, treating, pressurizing, and heating or chilling. In the end, our water system is about as energy consumptive as lighting. That means saving water saves energy. As an example, using drinking-quality "potable" water to irrigate lawns, golf courses, etc., is hugely wasteful when compared with watering those green spaces with raw water or treated wastewater that has a much lower energy requirement.

And in particular, saving on heated, treated water saves energy. These savings can be achieved through more efficient water heaters, only heating water when we need it, heating water with solar energy where appropriate, and using water-efficient dishwashers and clothes washers. The same concept is also true of saving food, reducing plastic waste, and reducing material usage in general.

## SILVER BULLET, SILVER BBS, OR CUSTOM BUCKSHOT

Most parties concerned with creating a more sustainable system of power production—one that will address climate change—understand that there is no one "silver bullet" that will solve the problem.

Every fuel and technology combination has limitations and downsides. Wind and solar energy may be nondepleting, cheap to operate, and free of emissions at the point of use, but they are not available 24 hours a day. Their energy density is low compared to fossil fuels, and at a utility scale, they can interfere with wildlife and other aspects of the environment. Nuclear energy is an energy-dense, carbon-free source, but the capital costs involved are high, and safety and waste disposal remain a problem. Coal is an inexpensive fuel source, but coal plants are some of the biggest greenhouse gas sources on the planet. Natural gas might be a good "bridge" fuel to renewable energy, but the gas system leaks methane, which is an even more potent greenhouse gas than carbon.

For these reasons, the silver bullet approach has been widely criticized, and you'll sometimes hear "there are no silver bullets, but lots of silver BBs" in the halls of Washington, DC. One former official commonly substituted the term "silver buckshot" for silver BBs.

The silver BB concept proclaims that there are myriad responses, technologies, and solutions that together can solve the world's energy problems. Advocates of the silver BB approach don't believe in preferential treatment of different energy resources or technologies, claiming that we need "all of the above." A systems problem needs a suite of solutions to be effective.

Perhaps the best answer is neither a silver bullet nor a silver BBs approach, but what we call a "custom buckshot" approach (stretching the munitions metaphor to its limit). In essence, there is no one-size-fits-all solution. Instead, it's vital that policy makers take into account nuances in both energy sources and regional variances in energy needs and renewable resources to maximize efficiency and carbon reduction gains. We need to match energy sources to workloads, optimizing the best uses for solar, wind, and nuclear as well as grid and decentralized power generation approaches. Each power source has characteristics that are best suited to particular energy loads, and properly matching them means greater efficiency on the whole.

In moving toward a more sustainable future of power production using a custom buckshot approach, it will be important to choose solutions with the greatest carbon return on investment (CROI). CROI refers to how much carbon dioxide or carbon dioxide equivalent emissions are removed from the atmosphere, or emissions avoided, for each dollar spent. Paying attention to maximizing CROI will give us the biggest bang for our environmental buck.

Key to getting the best CROI is recognizing regional differences, and it turns out that the best energy solutions are regional solutions. Consider the following example of energy solutions for three separate cities: Austin, Seattle, and Cleveland.

The cities of Austin and Seattle share many similarities. With 950,000 and 740,000 residents respectively, the cities are similar in size. While Austin proclaims itself the "Live Music Capital of the World," Seattle could easily make the same claim, as it's the birthplace of Jimi Hendrix, "grunge" music, and bands such as Nirvana, Pearl Jam, and Soundgarden. Both cities are largely sprawling and automobile-dependent. Both have been recognized as two of the "greenest" cities in the U.S. All similarities aside, however, their respective utilities' electric generation mix is very different.

The city of Austin meets its electric needs with almost 60 percent noncarbon emitting resources (nuclear, wind, and solar), about 25 percent coal, and 18 percent natural gas power. Thus, in Austin, focusing on overall reductions in electric usage wouldn't be nearly as beneficial as reductions in personal vehicle usage.

Seattle, on the other hand, meets its electricity with a mix of 90 percent hydropower, 4 percent wind, about 5 percent nuclear, and a little more than 1 percent fossil fuel. Therefore, Seattle City Light operates an almost carbon-neutral utility. Unlike Texas, hydropower resources are plentiful in the northwest United States. Despite its differences in primary energy generation sources, just like Austin, Seattle could make greater reductions in its overall carbon footprint by concentrating on reducing greenhouse gas emissions attributed to transportation than by reducing residential electric usage.

Then there's Cleveland. Calling itself the "Rock n' Roll Capital of the World," Cleveland is about the same size as Austin and Seattle. But unlike either of those cities, Cleveland gets 56 percent of its electricity from coal, 28 percent from nuclear, 11 percent from natural gas, and 5 percent from renewables. Cleveland is also subject to a humid climate, typical of much of the central U.S., with very warm, humid summers and cold, snowy winters. Consequently, residents in Cleveland use carbon-intense electricity to air condition in the summer and $CO_2$-producing natural gas to heat in the winter. In general, this means that significant improvements in residential weatherization would reap huge benefits. Austin, with its mild winters, seldom requires energy for residential heating, but has a high air conditioning load in the hot and long summers. Most of Seattle's electric load is for lighting during long, gray, but temperate winters.

So, despite the similarities of these three cities, the primary sources of carbon emissions are significantly different in each, and so, too, are the most economic options for avoiding carbon emissions. Austin, and even more so, Seattle, can make more meaningful reductions in overall carbon emissions avoided by developing policies that target reducing emissions from the transportation sector. In particular, electric vehicles in Seattle would be very low emissions. Cleveland can make greater reductions in carbon emissions, per avoided cost, through policies that reduce residential electric and gas consumption.

An easy way to see this approach is to take three sustainable energy efficiency technologies—energy-efficient light bulbs, electric vehicles, and a regular gasoline-battery hybrid vehicle such as the Prius—and see how they should be deployed to get the most greenhouse reductions in the least time at the least cost, which is the planning principle we have proposed. Based on the different electric utility composition of the three cities, it becomes clear that equal distribution of these three technologies would not achieve that goal.

We should change out all the light bulbs in Cleveland as soon as possible for the carbon savings, but changing the bulbs in Seattle would not save emissions. Placing an electric vehicle in Seattle would replace gasoline with clean hydropower, but we would be replacing gasoline with coal in Cleveland and a mix of fossil fuels and renewables in Austin. The third technology, a regular hybrid vehicle, would have similar savings in all three cities.

### Regional Solutions

Thus, getting to cleaner energy in the best possible way isn't a "do everything possible, everywhere possible" strategy, but should be made up of a patchwork of regional and seasonal strategies that take into account the various geographic differences in the available resources and the environmental impacts. For energy challenges and solutions, the "where" matters. The impacts vary by location and the solutions must vary too.

Once energy efficiency measures are implemented, it will be necessary to create utility-scale, regional renewable energy projects in concert with widespread, distributed renewable energy on smaller scales. After the best regional renewable energy resources are tapped, remaining energy needs should be met by grid-transmitted utility-scale renewables combined with energy storage or nuclear. All of this must happen in concert with the creation of a smart grid connected to new, powerful energy storage systems.

In the end, gains in efficiency and reforestation are the quickest and cheapest roads to greenhouse gas reductions and measures to address climate change, but energy efficiency will lose effectiveness in greenhouse gas reduction as the power on the grid becomes cleaner. So how will power production clean up its act, and what fuels and technologies will play the biggest roles in a more sustainable future for power production?

As we've said, solar has incredible potential. Part of the belief in the potential of solar comes from its obvious availability, and the fact that current and future technology allows any surface to become a solar power plant. Where we presently think of solar power coming from flat panels mounted in frames either on the ground or on rooftops, the next generation of flexible, modular solar panels, solar plastics, and solar paint will be attachable to almost any surface.

In the previous section we discussed the substantial potential for wind as a power source. Most scientists and the various wedge theories see wind playing a primary role in supplanting fossil fuels in a more sustainable power production future. Much of the optimism surrounding wind is rooted in the fact that it is, of course, all around us, everywhere. It's available both onshore, and increasingly, as the English, Germans, Japanese, and Dutch are showing, offshore. It is also very simple to convert flowing air into rotational motion to make electricity. Our more sustainable future will almost certainly see a huge growth in both large-scale wind power as well as the integration of low-power wind generation into our buildings. Small wind generators located on the top edges of tall buildings can capture the sometimes-intense winds created in the "urban canyons" of downtown environments.

Though nuclear power has its opponents, especially in the wake of the Fukushima disaster that prompted several countries to shut down many of their reactors, it should be part of the mix in a cleaner, more sustainable future for power production. As a more constant base load for sometimes intermittent solar and wind, and for high-density energy loads, emission free, nuclear power is an excellent fit.

While opponents point to the danger and waste associated with generating nuclear power, neither is as imminent as climate change, and both can be managed. Like solar and wind, nuclear power production will likely scale down in size in the years to come, from the huge, expensive reactors that are the norm today to small modular reactors (SMRs) about the size of a tractor-trailer. These will be discussed in the following chapter.

We can also expect the design of fourth-generation reactors to be much closer to fail-safe than the pressurized water reactors in use today, and probably with cleaner fuel sources and operations. Whether future nuclear power will be able to compete economically or receive adequate public acceptance remains to be seen. Nuclear might prove valuable to some niche applications, like the collaboration between Russia and China to develop floating reactors for the desalination of seawater and to meet energy requirements of the islands and reefs in the South China Sea.[15]

There is a current controversy over keeping nuclear power plants on-line, even though natural gas and renewables are pricing them out of the market in some states. Although there is a future sustainable scenario that only includes renewables and storage, without nuclear, we do not think this is the wisest course. Closing currently operating nuclear plants just eliminates a significant source of carbon-free electricity, and places additional burden on renewable resources.

Because we consider the climate problem to be a greater danger than dealing with spent nuclear fuel and safety issues, we think keeping current plants operating, even if requiring subsidy or regulatory implementation of carbon taxes, is a valid approach to addressing climate change. In general, fossil fuel plants need to be replaced first by renewables, and then nuclear where necessary.

As promising as both small modular reactors and fusion might appear, they both have one significant challenge: they fundamentally produce heat when what we want is electricity. Yes, we have many proven ways for turning that heat into electricity, but the construction, operation, and maintenance costs, along with energy conversion losses and environmental impact of the thermal energy to electricity conversion process, will always place these two technologies at a disadvantage to any electricity-generation technology, such as wind or solar, that convert a free energy source into electricity in one simple step. That is why we suggest that all energy needs should first be met by renewable energy flows and energy storage, and then use nuclear as necessary.

## Energy Storage

Almost all the energy and power needed by our cities, buildings, homes, and appliances has to be properly anticipated and generated for nearly immediate use. Any storage capability, whether at utility power plants or at distributed on-site sources, will make the overall system more efficient.

This is true whether the power comes from a small natural gas generator or from residential solar panels. As a greater percentage of utility scale power comes from solar and wind, energy storage will help to shift the time that these resources can be used. Thus, energy storage across all levels and capacities is critical for the full integration of renewable energy flows, and connected energy storage in utilities, buildings, and transportation is necessary to garner the largest efficiency gains possible in our more sustainable energy future. Proxies for storage, such as interruptible industrial processes and other demand response actions, will also be very valuable.

## CHAPTER 8 SUMMARY

1.  We can meet all our electricity needs sustainably through a combination of energy efficiency, renewable energy, energy storage, and nuclear.

2.  There is no single "silver bullet" solution for reducing greenhouse gas emissions. Different regions require different approaches.

3.  We should focus on converting 80 percent of our economy to emissions-free energy as soon as possible, since the last 20 percent is going to be more difficult and expensive.

4.  We have the technologies we need to transition to a sustainable economy.

Sam and Susan were concerned about the approaching tropical storm, but decided to go ahead with the party they had been planning for weeks. "Besides, we live in Houston. Gulf Coast storms come with the territory, like taco trucks and the Astros," Sam said.

Twenty-four hours later, they were glad they had continued, although the guests were fewer than they had hoped. The party had gone on, just as— sure enough—the storm had made landfall as a big, windy rainmaker. But although Houston's above-ground power grid, which was carrying power from wind and solar farms, had collapsed like a soggy spider web, underwater turbines out in the Gulf of Mexico kept the juice flowing.

Inside Sam and Susan's house, the lights didn't so much as flicker. The fact that their frozen margarita machine was being powered by undersea currents was just one of the many small miracles they had become accustomed to. The party went on without a hitch as the rain pounded down outside. The only downside was the ribs Sam had looked forward to barbecuing outdoors; it looked like frozen pizza was going to be the entrée d' deluge.

# ADVANCED POWER SYSTEMS

In the world of the not-too-distant future, humanity will power itself much differently than today. A hundred years from now, most of the energy mankind uses will be in the form of electricity, and it's likely that none of that electrical power will be generated from the burning of fossil fuels.

The world moves toward deep electrification that will be derived from energy sources both radical and familiar, far and near. Power production will be dominated by the harnessing of energy flows, including wind, solar, hydro, tidal, and the microharvesting of energy made possible by decades of developments in nanotechnology. Nuclear power generation will be used in many parts of the world for both base load electricity and "balancing" power for a renewable-energy-based grid. Each world region will move toward maximizing local and regional energy flows first, then choosing the best power source to meet specific base load or high-density power needs.

In the future, our renewably driven energy will be transmitted efficiently on both the micro and macro levels. Individual buildings, and even tiny, independent power sources, will be linked to continental and global power grids. High-efficiency transmission lines will let

Europe connect to African solar energy resources and let mainland China connect with wind energy sources in the mid-Pacific.

Complex prosumer relationships will form a highly integrated super-grid among buildings, transportation, and generation sources. The built infrastructure, including appliances, and the transportation sector will buy, sell, and store electricity. A massive marketplace of power will be mediated by synthetic intellects brought forth by the rise of artificial intelligence that enables rapid peer-to-peer transactions whereby a neighbor can sell electricity from the solar panels on their roof to their neighbor's house or the local power grid.

## Renewable Energy

Solar will ultimately be the primary energy harvested. In addition to the rooftop systems of today, we will see solar coverings providing massive amounts of energy as their use becomes ubiquitous on all manner of surfaces. Some even foresee solar power sources developed in concert with and in nature. Synthetic "solar trees," such as those in Singapore, could dot our landscape, intermingled with organic plants that may also be hosting integrated energy-harvesting technology. Landscapes may be offering energy, life-giving oxygen, and natural beauty in one hybrid ecosystem.

Others look to the Earth's great deserts—the Arabian, the Sahara, the Gobi, the Victoria, and the Mojave—as places where massive amounts of solar will be utilized for power production. Some even envision huge, replicating solar "breeder" plants producing new solar PV on-site, essentially growing the solar plant while simultaneously producing renewably sourced energy. Electric transmission lines using advanced super-conducting materials could connect the solar power potential of the Sahara in North Africa to the European continent and the rest of Africa.

According to Nobel Laureate in physics, Robert B. Laughlin, Americans may be the first to experience the benefits of desert solar power on a metropolitan scale. In *Powering the Future*, Laughlin points out that while all the great deserts receive copious amounts of sunlight, "...the Mojave is the only one also endowed with lots of power lines and close proximity to large, affluent metropolitan areas. Those things are just as important as sunshine when one is building power plants. Accordingly, the first great world city likely to become solar powered when fossil fuels begin to flag (if any do) is not Cairo or Karachi but rather Los Angeles."[1]

Others look to the skies as a large-scale, alternative-energy generation source for the distant future. Wind farms already generate electricity off shore today. Larger, more efficient complexes, both on and offshore, might be joined by even more diverse and distributed wind-capturing approaches. Open-ocean wind farms could generate much more power than the current land-based or even coastal offshore wind resources. Whereas land-based farms may be limited to approximately 1.5 watts per square meter, one study has shown that a wind farm located in the North Atlantic Ocean could generate more than 6 watts per square meter.[2]

Of course, the largest features on planet Earth are our oceans. As of today, we've barely begun to harness any of the power present in their energy flows. We do already have a small number of machines generating electricity from tides and waves. The distant future will almost certainly be a place where we convert some of the enormous power contained in the motion of the oceans, both tides and waves, as well as the motions interior to the ocean such as ocean currents.

Today there are a half-dozen prototype devices anchored to the ocean floor converting the energy of ocean tidal and current flows into electricity. Orbital Marine Power recently completed a two-year demonstration of a full-scale, 2MW tidal generator off Scotland's Orkney islands.[3] The technology is now being deployed in the form of several 2MW devices off the western tip of the Hebridean island Islay in a tidal energy farm that began construction in 2019. The farm is expanding up to 400MW by 2024.[4]

Unlike wind turbines, ocean tidal and current turbines are out of sight on the ocean floor. And since the energy and power density of flowing water is almost 1000 times greater than that of air, the area required to generate equivalent levels of power is vastly smaller. Also, tidal and current flows are very predictable. In the case of ocean currents, the flow and power production are 24/7 without any interruption. Just imagine powering the entire Eastern coast of the United States with an array of seabed power turbines stretching up the Gulf Stream from the Florida Keys to Virginia.

Laughlin notes that as the use of fossil fuels begins to wane, the deep ocean will be increasingly important to mankind's power production, and will be utilized for things like gathering deepwater methane, accessing geothermal energy, and storing energy in pressure and thermal gradients. Laughlin predicts this undersea era of power production and storage will begin near the turn of the next century.[5] Much of the work required to ply energy from such inhospitable depths, he points out, will be performed by remotely operated vehicles—robots—much like the ones Bob Ballard used to locate and photograph the wreck of the Titanic.[6]

### Small Modular Reactors

The future of nuclear power will probably come in one of two novel forms. The first are small modular reactors (SMRs), which could be small enough to be transported on the back of a tractor-trailer and be an addition to the already expanding distributed energy grid. Smaller reactors may better fit niche, high-power applications and may be easier to manage through construction and operation.

Ontario Power Generation plans to meet a forecast energy supply gap in the 2030s using SMRs. They are teaming up with Saskatchewan province, a major uranium mining area, for a fleet of SMRs across Canada. The Canadian SMR company Terrestrial Energy Inc. has a design for a small modular Integral Molten Salt Reactor (IMSR) that seems ideal. They claim it can be constructed in less than four years and can be trucked to the site. It also uses molten salt for a passive heat removal system that makes it "walk-away safe."[7]

Russia deployed its first commercial floating nuclear reactor in 2016. The 70 MW power plant was designed on the basis of their existing nuclear reactors for icebreaker ships, and will be used for power, heat, and desalinization in the Siberia region. Several countries are interested in such small floating reactors to supply power for everything from seawater desalination to oil and gas exploration.[8]

As SMRs are smaller than current commercial reactors, they could be manufactured at scale in factories, which means they could be produced along a declining cost curve similar to what solar cell manufacturing experienced as solar adoptions increased. Getting away from massive nuclear power plants built one-at-a-time on-site to mass production in factories could be a key enabler in making nuclear power more affordable. And like their GW-scale brethren, SMRs could provide the 24/7, emissions-free electricity necessary to shore up a mostly renewable-energy grid. Such an approach would minimize the huge amounts of electrical energy storage required to meet extended periods of low renewables production.

## Fusion

Others are still looking to fusion to provide the holy grail of nuclear power. We know fusion works, as demonstrated by the sun, but making it work in an energy-productive way on earth is vexing. Researchers at the National Ignition Facility in California completed the first fusion experiment ever to garner more energy than was expended.[9] Researchers from the European Union are hopeful that they'll be able to build a prototype fusion reactor before 2050.[10] The International Thermonuclear Experimental Reactor (ITER) in southern France is funded by an international coalition of 35 countries, but it is billions over budget and years behind schedule.[11]

A quicker solution may come from private sector efforts at smaller scales. Lockheed Martin has received a patent for a compact fusion reactor that could be as small as a shipping container, but could power 80,000 homes.[12] A handful of other private start-ups are also promising fusion power in the next couple of decades. The MIT spinout startup, Commonwealth Fusion Systems, plans to have a working reactor in the next 15 years.[13]

LPP Fusion achieved a world-record temperature with their Focus Fusion-1, a device that fits in a small room,[14] and Helion Energy is working on a container-sized, 50 MW fusion reactor.[15]

Fusion power always seems to be the power of the future, but if commercially viable fusion power can be achieved, it will be a limitless source of power that emits neither greenhouse gases nor long-lived radioactive wastes, and without the danger of a catastrophic meltdown. Of course, fusion still is just a manner to create heat to convert to electricity. Therefore it could still struggle to compete economically with solar and wind devices that generate electricity directly from free energy flows.

## Micro energy harvesting

On another level, we could micro harvest even smaller energy flows to produce power in the future. Today, batteries power most small electronic devices. But there are already micro energy harvesting devices on the market that can harvest small energy flows like thermal gradients and air flows that can, for some applications, replace batteries.[16] While experts don't see such small, captured energy being useful on a utility scale, many envision it wirelessly powering the sensors and other micro devices that will be soon be all around us. Thermoelectric, piezo-electric, fabric, and wearable-based energy devices will find widespread use in powering sensors and meeting the "vampire" loads of charging small electric devices. This is a much better solution than trying to "hardwire" millions of sensors in our buildings.

We might meet some of our future energy needs from evaporation on our lakes and reservoirs. Biophysicist Ozgur Sahin has found a way to use spore-coated plastic strips floating on water to elongate and shorten in response to evaporation. He then connects the device to a generator that converts the motion into electricity. Although the power of an individual device is small, the overall potential could be significant, and the evaporation device operates at night as well as day.[17]

Prospective power sources are all about us, and accessing them for electrical generation is simply a matter of the right technology. One such example is the Soccket soccer ball, which can harness some of the energy of its motion during play, and then be used to power small devices such as lamps.[18] All manner of similar devices could capture motion flows in the years to come.

## Grids

Clean manufacturing, transmission lines made from exotic materials, and widespread use of ultrastrong, lightweight materials will make power production and transmission increasingly efficient.

As the production of electricity and its transmission become more efficient, they'll grow into "intelligent super grids." Buckminster Fuller once said, "We must integrate the world's electrical-energy networks. We must be able to continually integrate the progressive night-into-day and day-into-night hemispheres of our revolving planet. With all the world's electric energy needs being supplied by a twenty-four-hour-around, omni-integrated network, all of yesterday's, one-half-the-time-unemployed, standby generators will be usable all the time, thus swiftly doubling the operating capacity of the world's electrical energy grid."[19] What may have seemed a futuristic hope will almost certainly become commonplace reality as continental and international supergrids replace the piecemeal network of today.

Carbon nanotubes and other exotic nano materials promise to produce extremely strong and very low resistance transmission lines. In just the last few decades, we have fielded a massive global information network of fiber optic cables. With nearly lossless superconducting materials, it might be possible to replicate the global information network with a global energy network and bring Fuller's vision into a global renewable energy reality.

Such a network would negate the need for much of the energy storage capacity we are so focused on building today. A global network would allow us to nearly instantly move electrical energy from anywhere in the world it is being generated to anywhere it is needed. Such a global network would first be built to interconnect vast contiguous regions and then continents. We know that, with today's technology, a solar PV farm of 4,300 square miles (or 1 percent of the Sahara) could produce the equivalent of all global energy use.[20] If we were to build several solar PV farms a fraction of that size and strategically place them in the major global deserts, we could interconnect them and power the planet.

The role of power grids will also change. No longer will they be passive transmission and distribution systems, they will be heavily involved in the coordination and integration of an increasing number of suppliers and the distributed generation system as a whole. Grids of the future will efficiently move energy among generation, storage, and consumption sites. Artificial intelligence systems will broker energy sales to and from innumerable disparate sources. They'll monitor, control, and protect the quality of the power flow and will be fully automated at a level that is difficult to comprehend today.

## Sentient-Appearing Power Systems

In all likelihood, global supergrids of the far future will be self-controlling and need little human input day-to-day. They'll be self-healing, self-sufficient, and have sentient-appearing interfaces. They'll monitor themselves, solve problems before they happen, and interact with the built infrastructure and global transportation systems in ways that will make the entirety of energy production and consumption as efficient as possible. Operational and market decisions will be made by automated systems. Power plants are already highly automated, and operators are often just watching the load and generation responses between the buildings and the power plants.

## Autonomy

Furthermore, the manufacture of power production equipment, including renewables, will increasingly be automated. New cost reductions in solar would be expected from the large-scale introduction of robotics into production.

Robots will also take over more inspection and maintenance duties in the industry. Tiny robots may perform wind turbine blade inspections, and robots and drones will be responsible for checking transmission lines. Robots will install and control solar PV as well.

Likewise, a "self-healing" grid is emerging, where sensors, robotic inspectors, and computers detect and correct potential problems, as well as automatically rerouting energy and correcting faults as they occur.

Finally, it is possible that robots will autonomously acquire and process the primary fuels for our energy systems. It is easy to imagine automated systems locating, mining, and processing fossil fuels; setting up solar and wind farms; drilling geothermal wells; and all the other tasks necessary to convert primary energy sources into electricity and dispatch it to the grid.

This system, as with the building and transportation systems discussed earlier, could be controlled and operated by artificial intelligence systems that would interact with us as if they were sentient. Our energy systems would certainly form working relationships with the buildings, transportation systems, cities, and other entities that they power.

## CHAPTER 9 SUMMARY

1.  Our power systems will continue to get smarter and more efficient as they incorporate artificial intelligence and become more integrated.

2.  Regional power systems will harvest renewable energy flows, backed by nuclear and energy storage.

3.  There will be large-scale implementation of small-scale power technologies.

4.  The global grid will move power among continents.

5.  The oceans and deserts will be a major source of power and materials.

# PART 5

CHAPTER

# 10

# OUR CRYSTAL BALL

It is time now for our concluding thoughts on the future of buildings, transportation, and power. We will revisit the Energy Efficiency Megatrend and speculate on further technologies. We will also address the question of whether we can attain a sustainable, clean future for these sectors. And we will highlight some obstacles to the sustainable and advanced-technology scenarios.

## ENERGY EFFICIENCY MEGATREND

The Energy Efficiency Megatrend is the principle that technology continually evolves to make energy transformations more efficient. In chapter 1 we discussed three consequences of that trend for future technology. Let's return now to look at those three and speculate on some areas of technology that may further evolve.

The first consequence was that the three economic sectors addressed—buildings, transportation, and power—would meet their functions with less material, motion, and time.

We have seen the development of advanced materials in all three sectors, especially new materials based on nanotechnology. We have also seen an increase in speed, from the time needed to construct a building to supersonic aircraft and hyperloop trains.

Nanotechnology will continue to be the driver in material technology, creating new materials, manufacturing tools, and products to atomic specifications while improving quality and reducing manufacturing waste.

Nanotechnology will be especially important in advanced power systems. This level of technology will allow us to double the efficiency of solar panels. It could significantly increase battery storage capacity and create superconductors for almost effortless transmission of power.

The second consequence was the development of sentient-appearing interfaces for buildings and vehicles. We are getting used to speaking out loud and making gestures to give commands, ask questions, or otherwise interact with machines.

While we are just starting to be comfortable talking to our things, we may be able to dispense with the conversation. Many companies are now working on brain-computer interfaces, and it is easy to imagine a future where we converse with a building or vehicle "in our heads," giving directives and changing things by just thinking or visualizing what we want. This can be seen as a form of technological telekinesis. If preset automated systems can be activated by thought, then we could accomplish many tasks within our buildings and control vehicles with our minds. In fact, the first transportation brain-machine control has already been demonstrated. A quadriplegic has been able to walk using an exoskeleton controlled by his thoughts.[1]

We might be able to use the sensors and cameras on buildings and vehicles to expand our senses, seeing and "sensing" their environment and tapping into their information and intelligence.

As we start to integrate with our technological environment, the issues of privacy, public access, equality, human and machine rights, and even identity will become more pronounced. All the issues raised by artificial intelligence and posthuman evolution will come to the fore.

When buildings and transportation systems are able to recognize and react to us, then we will be living in a "magical" urban environment that is always aware of us and what we are doing. The artificial intelligence systems will enhance that awareness beyond any surveillance being done in the world today.

This is certainly good for emergency response and customer service, and it could certainly prevent a lot of crime. But privacy will become difficult and more expensive.

The third consequence of the Energy Efficiency Megatrend was that buildings, transportation systems, and power systems are forming a single, interdependent, thermodynamic system. Electrical power generation is being physically distributed throughout the system, and power flows can be consumed, stored, or moved. This means that our overall power system should become more powerful and reliable over time.

## SUSTAINABLE FUTURE

There is no doubt that we can achieve sustainable, emissions-free operations of our buildings, transportation, and power. This book has shown the many ways we are moving toward that sustainable future.

Some parts of this transition are going to be relatively fast and cheap. The generation of electricity will probably be the easiest to convert to renewable energy and nuclear resources. The two main sources of renewable energy, wind and solar, can be expected to compete successfully against oil, coal, and even natural gas. Combined with government mandates and incentives, we could rapidly transform the power sector.

Electrification of the urban transportation sector should also be a bright spot. Electric vehicles will soon reach costs comparable with conventional cars and they are already cheaper to operate. In ten years, automakers will probably not be selling vehicles that cannot be plugged in. Eventually, all movement of people and goods within our large cities will be cleaner, quieter, and swifter. Furthermore, parts of the transportation sector should also see significant cost reductions when the transition to electric motors becomes complete.

However, there are parts of the economy, including significant areas of the transportation sector, where replacing fossil fuels with renewable energy is going to be difficult and unlikely to be cheaper. Aviation, international shipping, high-temperature manufacturing, and some other areas will be very difficult to wean off fossil fuels.

For example, the global movement of freight and people will be difficult to change. International shipping is powered by two technologies: the diesel engine and the gas turbine.[2] These two technologies are not very amenable to substitution by an electric motor in those markets.

It is possible to develop renewable energy-based synthetic hydrocarbons to do the job, and atmospheric carbon could be mined and turned into fuel, but the fuel will not be cheap in the early phases of deploying that technology. It is hard to imagine a renewable energy-based fuel that could compete economically with petroleum for these segments of the economy. Therefore, that transformation will have to be driven by policy, against strong market headwinds.

And there will still be some applications where it will make more sense to capture and store emissions than change the fuel.

We do think nuclear will be part of the energy solution, although the vast preponderance of energy needs will be met through harvesting regional renewable energy flows. Still, nuclear can often be the cleanest solution to some high density and nonintermittent power needs.

Movement to a sustainable economy means adjusting to the timing of renewable energy flows. That may mean both a shift in the time we do certain tasks, and some travel times may take longer. We may need to sacrifice speed for sustainability in parts of our lives.

Just as today, we do not expect advanced technologies to be available to everyone. Some of the future energy technologies referenced earlier may only be enjoyed by the wealthy "one percent," while other technologies, such as solar photovoltaics, will enable cheap and abundant energy throughout the world.

Technological advances will also spawn social, cultural, and legal issues. Desynchronization will show up everywhere as our technology advances outpace the ability of our business models, government policies, social norms, and legal systems to keep up.

We should also not harbor the illusion that a rapid transition to a more sustainable electricity generation and transportation system will be environmentally benign. Every technology has environmental impact.

Certain regions will increase mining and suffer environmental damage for the rare earth metals and other materials that support clean energy technology. Wind and solar farms do impact the local environment, and can harm local wildlife. The manufacturing of electric vehicles is more polluting than making conventional cars. Used batteries and renewable energy equipment will raise a substantial waste and recycling issue. These environmental problems need to be mitigated as much as possible, but addressing the climate crises will involve such trade-offs.

## BUMPS IN THE ROAD

### Embedded Energy

Embedded energy is the sum of all the energy used to produce something: in our case, buildings, vehicles, and power systems. Energy is required to manufacture and install wind turbines, solar panels, batteries, and electric vehicles. The irony is that we must burn fossil fuels to manufacture the renewable energy technology we need to generate electricity without burning fossil fuels. A lot of steel, concrete, and advanced plastics will be used to transition to a sustainable economy.

Consider the fossil fuel requirements of a single 5 MW wind turbine. The wind turbine is basically a tall steel tower, sitting on a concrete base, with long fiberglass blades. The steel alone averages 150 metric tons for the foundations, 250 metric tons for the rotor hubs and gearbox and generator housing, and 500 metric tons for the towers. Famed energy historian Vaclav Smil has calculated that if we wanted to meet 25 percent of forecasted global energy demand in 2030 with wind turbines, we would need the equivalent of more than 600 million metric tons of coal to make the steel and concrete. While those numbers are jaw-dropping, they do not anticipate the significant drops in material requirements for the latest turbines. Swedish company Modvion has developed a wind turbine tower from composite wood, which will soon be tested by the Sweden Wind Power Technology Centre. The company claims that a full sized wooden turbine mast would avoid 2,000 tons of carbon emissions produced by a steel tower.[3]

The fiberglass wind turbine blades, along with other advanced plastics, are a solely petrochemical product. When you consider the other elements of the wind turbine that contain epoxy or polyester resins, glass and fiber-reinforced composites, you would need an additional 90 million metric tons of crude oil.[4]

There is no doubt that a good wind turbine would generate as much energy as it has embodied in a year or so, and then generate emission-free electricity for decades. But that does not eliminate the fossil fuels to get there in the first place. Recall that the wind, water, and solar plan advocated earlier calls for 3.8 million 5 MW turbines to meet 50 percent of our load—double what Smil calculated.

Similar problems are involved with solar cell manufacturing, geothermal plants, building new dams (involving a lot of concrete), and most other forms of renewable energy power generation.

Cement and steel are particularly fossil-fuel-intensive industries. In 2016, about 8 percent of global carbon dioxide emissions came from the cement industry. The steel industry also generates between 7 and 9 percent of global carbon dioxide emissions.[5] These industries produce such high emissions because of the heat necessary for production, the chemical reactions in production, and the fossil fuel consumption in mining and transportation of the raw materials and finished products.

Cement kilns need to be kept at 2,640°F.[6] The main ingredient of steel, pig iron, is produced in a blast furnace. The production of glass and aluminum, and even the recycling of aluminum and steel, require high temperatures most economically achieved through combusting fossil fuels.

And then there are the chemical reactions involved. Smelting iron ore for steel produces carbon dioxide. About half of the emissions from making cement come from converting limestone into lime, which also produces carbon dioxide.[7]

The decarbonization of cement and steel production will be extremely difficult. It will be very difficult to change the manufacturing process, and it may not be possible to change some parts.

Richard Heinberg and David Fridley at the Post Carbon Institute reviewed the options to obtain such high temperatures without burning fuels and found them lacking. Using solar or wind to produce heat from electricity would be two to three times more costly than burning coal or gas, though the inclusion of electric arc heaters in steel mills has been revolutionary. Thankfully, hydrogen fuels can be used to reduce emissions and improve efficiency for both industries.

Solar thermal could produce temperatures high enough to manufacture steel or cement, but would require focusing optics, which poses a challenge to integrate it at an industrial scale with cement kilns and blast furnaces. It is unclear how the heat could be expanded to a controllable system, and we do not have ways to store large volumes of high-temperature heat necessary for metal smelting and cement making.

Using biomass and biogas for such manufacturing could decimate our forests or exceed the available organic waste streams. Using hydrogen is theoretically possible, but would require a massive redesign of high-temperature industrial processes and hydrogen production is energy-intensive on its own.[8] Even the substitution of new, lighter, and stronger materials like carbon fiber require high-temperature manufacturing.

The energy necessary to manufacture and install the transportation infrastructure—electric vehicles, planes, ships, roads, bridges, etc.—will also be much more difficult to produce from renewable energy resources than fossil fuels.

However, embedded energy should not be used as an excuse to avoid switching to renewable energy sources. The amount of fossil fuels necessary to manufacture sustainable transportation, buildings, and power generation is a fraction of the fossil fuels we currently use in these sectors. We can mostly eliminate the burning of fossil fuels in our economy while dedicating the necessary fossil fuels to make the renewable energy technology we need. This should be considered when setting goals such as a 100 percent renewable energy economy. We cannot have a goal of 100 percent carbon-free manufacturing at the same time we are working toward carbon-free electrical generation, buildings, and transportation.

We certainly can continue to innovate and redesign manufacturing to be less carbon intensive. Current efforts to produce "green" concrete include substitution of fly ash and waste for a portion of the concrete mix, recycling concrete, and developing cement replacement materials.[9] Nanotechnology may provide ways to make concrete at lower temperatures, which would be a major innovation.

But basic redesign of these core industries will take time, and the need for renewable energy and transportation technology is urgent. We must not be naive about the necessity of fossil fuels to manufacture the technology.

### Critical Elements and Materials

The use of certain critical elements and material in our modern technology could be a limiting factor in our energy future. The EU has listed and tracks sources and supplies of 50 critical raw materials, including several rare earth elements (REEs) that are key in the manufacturing of wind turbines, solar panels, electric vehicles, and batteries.[10] The use of REEs such as neodymium, dysprosium, and lithium could be a limiting factor to the rate at which we can rapidly scale production of clean energy technology without running the risk of short-term depletion and rapid price escalation.

The problem is not the existence of these REEs and raw materials, as they are actually not rare at all. They are quite abundant and distributed throughout our world, but they usually occur in very small concentrations. The problem is that they appear in only a few places that are economically viable to mine today—primarily countries with lax environmental, safety, and health laws, and low labor costs. For instance, 60 percent of the cobalt being used in electric vehicle batteries today comes from one country: the Democratic Republic of the Congo.[11]

This means that obtaining greater quantities of these materials from a broader and more sustainable range of resources could be expensive. And unfortunately, it cannot be done quickly. A new mine usually takes 10 to 12 years to bring to full production.[12] A sudden surge in demand resulting from government action could not be met by an immediate increase in production. We will need to make the clean energy transition necessary over the next decade with mostly our current mining capacity.

Recycling is not profitable yet, but higher prices will change that. This "urban mining" can become a major resource for a few elements. But even that will be slow and difficult to turn into a major supply resource.

There certainly are innovative efforts underway to replace the use of rare earth elements with new processes or alternate materials. 3D printing of vehicle chassis is one example.

The modern smartphone contains the majority of the periodic table.[13] It remains to be seen if new and more environmentally friendly sources of these materials and/or alternatives can be developed at competitive prices. The bottom line is that supply scarcity and higher prices for REE could be a damping force early in the transition to a cleaner energy future.

## Electricity

Our "smart" future is dependent on electricity generation. Since the 19th century, the increasing information intensity of our technology has been enabled by electricity, from lights and computers to sensors and communication technology.

Although electric service will be more reliable in many ways through redundancy and distributed power and storage, there will still be outages, and the impact will be greater and greater. We have seen the disruption that massive blackouts can cause. If we electrify the transportation system, that shuts down as well.

Physical attacks, natural disasters, cyber attacks, or other sources of disruption can wreak havoc. Climate change may increase the frequency and intensity of extreme weather events. Floods, wildfires, hurricanes, and winter storms can take out large portions of the electrical system. Microgrids will flourish, but they also are vulnerable to extreme weather that directly hits them.

Electricity is also the lifeblood of artificial intelligence. Sentient-appearing buildings and vehicles become dumb immediately when the flow of electrons stops. It will be very jarring to be living in a sentient-appearing environment, where you are used to interacting with the Internet of Things, and suddenly have everything become dumb.

The future may be bright, but it is not invulnerable. As a friend and former mayor of Austin, Lee Leffingwell, noted: "If it runs on electricity, sooner or later it will let you down."

## *Complexity*

Utopian future scenarios always presume that everything works. It would be easy to envision a wonderful future where all of this intelligent technology works seamlessly and easily to provide us with a world of plenty, where nothing breaks down.

Our technological history has not produced that, although it is certainly the case that modern technology works with a remarkably high degree of accuracy and reliability. The problem is that we are increasing the complexity of our systems by orders of magnitude.

Sensors, for example, are going to be everywhere. If even a small percentage of sensors are not working correctly, there will be a lot of false signals. Sensor failure is not just annoying; the cause of several airplane crashes came down to sensors and the automated systems response to those sensors.

When you add bugs in the program, there is a lot to go wrong. Even if future AI systems start to find and fix problems as they arise, getting from here to there could be frustrating.

It already seems like we spend a lot of our time dealing with our technology not working just right—smartphones, cable services, wireless connections, computers, and other high-tech equipment. "Reboot" has become a common solution to make all sorts of things work.

And this is not even considering the criminal or enemy activities on the web. In our youth, "identity theft" was just a fun plot line for mystery fiction writers. Now, cybercrime is a major international criminal enterprise.

"Smart" cities are pitched as a solution to many of our urban problems. But there is a problem with the lifespan of new smart technology in comparison to urban public works projects designed to last for decades. Are city leaders going to ask us to pay the cost to keep up with the latest smart technology? Are they going to be able to afford the cost of maintaining all the sensors, hardware, and software required to keep a smart city smart?

There is a strong argument for urban environments to develop both "smart" and "dumb" operations together, so basic city functions can be met in the event of electrical outages, sensor failures, cyber attacks, and other breakdowns.

## CONCLUSION

We have described a future where our buildings, transportation systems, and power grids become efficient, intelligent and aware, where buildings and vehicles talk among themselves and to us, helping in our daily lives.

We think there is a good chance our future infrastructure will be clean and provide sustainable energy for our needs. We should be assured that we are moving in the right direction, and daily make progress. As to how long it takes to get there, and all the problems associated with the journey, we will guess along with the rest.

# ACKNOWLEDGMENTS

Many people and institutions helped us in the formulation of this book and the actual writing and production. We first want to thank those who helped us put the book together, and then we each want to thank those who influenced and inspired the ideas in the book.

For the actual production of this book, we will start with our editor, Lori Handelman of Clear Voice Editing. Her suggestions and critical eye turned the book from a dry statistical work to an enjoyable read.

We also thank Jeff Phillips for the graphs, and for his suggestions about how to illustrate some of the data. A special thanks to John Wilson for the "Our Energy Future" illustration.

The endnotes were skillfully handled by Ashkan Jahangeri, Madeleine Detelich, Dr. Isabella Gee, Kayla Fenton, and Brittany Speetles. The writing skills and help of Chris Maddock were invaluable in creating the first draft. John T. Davis drafted the wonderful vignettes, first suggested by Michael Osborne. Many researchers and fact-checkers contributed, including Beth Martinez, Jonathan Valdez, Jorge Navas, Kevin Reed, Berkeley Beauchot, Maggie Duffy, Mark Reid, Reuben Gol, Somtoochukwu Ethel Ik-Ejiofor, and Whitney Zimmerman.

## ROGER:

Numerous teachers and colleagues over the years informed me and opened my eyes to the technology trends in buildings and transportation from the city government view. Special thanks to my coworkers in the electric utility industry, many of whom were leaders in the clean energy trends discussed in the power sector.

The Energy Institute of the University of Texas generously funded the time and resources necessary for me to produce this work. I want to give special thanks to the Environmental Defense Fund, which provided grants for some of my work. EDF has been a source of innovation in the clean energy field for a long time. Particular thanks to Jim Marston for his support at EDF.

The chapter on sustainable building was particularly influenced by Amory Lovins, whose work is heavily cited. I was fortunate to collaborate with Amory early in my career, and his visionary leadership in energy efficiency has been inspirational throughout my life. Also, the opportunity to work over many years with the Green Building program in Austin, one of the first in the nation, was invaluable to the buildings section. They have been at the cutting edge of building technology for the past three decades and over the years have provided me with many insights into sustainable building.

The transportation section owes much to the people I worked with in the electric vehicle renaissance, particularly Michael Osborne, Tom "Smitty" Smith, and the early leaders in the plug-in hybrid movement. Special thanks to Madeline Detelich, whose research and drafting helped me rewrite and update the two transportation chapters. Thanks to Dr. Isabella Gee for her research on aviation biofuels. Finally, Dr. Dave Tuttle provided both technical insights and practical wisdom into the electrification of the transportation sector.

Dr. Fred Beach updated and redrafted the three chapters on power. Without his overall grasp on our energy systems and knowledge of current technological development, I doubt this book would have been finished. Dr. Beach and I were the instructors for a Policy Research Project at the LBJ School of Public Affairs, where the students very effectively analyzed the various business models being tested in the electric utility industry and informed the business model section of the book.

I give a special thanks to Dr. Stephen Suttle, Dr. Roger Gammon, PA JoAnna Brand, and all the great nurses and staff at Austin Heart Hospital, which provided me the opportunity to finish this work.

I also thank my coauthor, Michael. He has been a mentor in the ways of the academic world, and has provided many hours of intellectual enjoyment as we thought though the future of technology. His deep knowledge of energy and engineering were invaluable in shaping my vision of the future.

Finally, I thank my wife Jo, not only for her patience and encouragement, but also for her writing and editing skills that helped shape the book.

## MICHAEL:

In addition to all the people Roger thanked, I also acknowledge my many research collaborators, coauthors, and students over the last 15-plus years. I learned many key insights from them, some of which made their way into this book. I thank Ms. Marianne Shivers Gonzalez, who coauthored an article with Roger and me in the Winter 2013 edition of *Issues in Science and Technology*, which helped us organize our thinking about the slow pace of innovation in energy technologies. Marianne helped us significantly for several years, from around 2011 and beyond with writing, research, editing, and organization of our various thoughts.

I also thank CLEE (the Center for Lifelong Engineering Education), who gave us a chance to teach a one-day short course on the subject of the book in Spring 2013 to an audience of industry professionals. This course gave us an opportunity to organize our thinking and to test it out on a mix of experts and enthusiasts. The preparatory work for that course helped us move forward with the book.

I especially thank Roger. This book was born from his unique insights, and I was honored he invited me to join the project. He taught me a lot and was always a pleasure to work with. We should all hope for collaborators as smart, gracious, and generous.

# NOTES

## INTRODUCTION

1. Michael E. Webber, Roger D. Duncan, and Marianne Shivers Gonzalez, "Four Technologies and a Conundrum: The Glacial Pace of Energy Innovation," *Issues in Science and Technology* 29, no. 2 (Winter 2013): 79–84.

2. International Energy Association, "Mission."

3. Energy Information Administration, "Strategic Plan."

4. The United States Department of Energy defines its "reference case" as one comprising "current laws and regulations," assuming that "existing laws and regulations will remain unchanged throughout the projection period, unless the legislation establishing them sets a sunset date or specifies how they will change." *Annual Energy Outlook 2012*, U.S. Energy Information Agency, 18.

   The International Energy Agency developed two scenarios that we use. The Current Policies scenario is defined as "Government policies that had been enacted or adopted" as of the date of the report, and excludes ambitions and targets. The New Policies scenario is "Existing policies are maintained and recently announced commitments and plans, including those yet to be formally adopted, are implemented in a cautious manner." Note that the New Policies scenario goes beyond the EIA reference case in that it assumes that governments will achieve announced commitments before the relevant laws and regulations have been developed. IEA WEO, 2012, 35.

5. The International Energy Agency has developed two scenarios specifically addressing a sustainable future: the 450 scenario and the Efficient World scenario. The 450 scenario identifies an approach that limits the concentration of carbon dioxide in the atmosphere to 450 parts per million or less. This case assumes that "Policies are adopted that put the world on a pathway that is consistent with having around a 50% chance of limiting the global increase in average temperature to two degrees C in the long term, compared with pre-industrial levels." The Efficient World scenario assumes that "all energy efficient investments that are economically viable are made and all necessary policies to eliminate market barriers to energy efficiency are adopted." IEA WEO, 2012, 35.

6. Arthur C. Clarke, *Profiles of the Future: An Inquiry into the Limits of the Possible,* New York: Popular Library, 1973.

## CHAPTER 1: TECHNOLOGY MEGATRENDS

1. R. Buckminster Fuller, *Nine Chains to the Moon,* Carbondale: Southern Illinois University Press, 1938, 253.

2. R. Buckminster Fuller, personal conversation with author Roger Duncan, May 1981.

3. Ray Kurzweil, *The Age of Spiritual Machines: When Computers Exceed Human Intelligence,* New York: The Penguin Group, 1999, 30.

4. Gordon E. Moore, "Cramming more components onto integrated circuits," *Electronics Magazine,* April 19, 1965.

5. Ray Kurzweil, *The Singularity Is Near,* New York City: Viking Press, 2005, 67.

6. Ibid. 243–244.

7. Robert Bryce, *Smaller Faster Lighter Denser Cheaper: How Innovation Keeps Proving the Catastrophists Wrong,* New York: Public Affairs, 2014.

8. James Canton, *Future Smart: Managing the Game-Changing Trends that Will Transform Your World*, Cambridge: Da Capo Press, 2015, 99.

9. Richard Munson, *From Edison to Enron*, Westport: Praeger, 2005, 53–54.

10. Alexis Madrigal, *Powering the Dream*, Cambridge: Da Capo Press, 2011, 39.

11. General Electric, "Becoming Brilliant: Fly Into the Future With GE's Brilliant Machines," Advertisement.

12. Carlos Madrid, "Galleon Victoria: First Ship to Circumnavigate the World," Guampedia.

13. "Douglas World Cruiser *Chicago*," Smithsonian National Air and Space Museum.

14. Robert Braeunig, "Vostok," Rocket and Space Technology.

15. Kevin Kelly, *What Technology Wants*, New York: Penguin Group, 2010.

16. Ibid. 166.

17. Ibid. 166–167.

18. Webber, Duncan, and Gonzalez, "Four Technologies and a Conundrum," 79–84.

19. Ibid.

20. Vaclav Smil, "Moore's Curse," *IEEE Spectrum*, March 19, 2015.

21. Ibid.

22. Vaclav Smil, *Energy Transitions*, Santa Barbara: Praeger, 2010, 123.

23. Kelly Pickerel, "The long history of solar PV," Solar Power World.

24. John F. Geisz et al., "Six-junction III-V solar cells with d47.1% conversion efficiency under 143Suns concentration", Nature Energy, 2020.

25. Vaclav Smil, *Creating the Twentieth Century: Technical Innovations of 1867–1914 and Their Lasting Impact,* New York: Oxford University Press, 2005.

26. Ibid.

27. Alvin Toffler and Heidi Toffler, *Revolutionary Wealth,* New York: Knopf, 2006, 31.

28. Smil, "Moore's Curse."

29. David Kline, "The Embedded Internet." *Wired,* October 1, 1996.

30. "Your car has more computing power than the system that guided Apollo astronauts to the moon," Physics.org.

31. There are three supportive sources, listed below:

    a. Asha Sharma et al., "A carbon nanotube optical rectenna," *Nature Nanotechnology Letters,* September 28, 2015.

    b. Jeong Hee Shin et al., "Ultrafast metal-insulator-multi-wall carbon nanotube tunneling diode employing asymmetrical structure effect," *Carbon 102,* 2016.

    c. Naser Sedghi et al., "Towards Rectennas for Solar Energy Harvesting," *IEEE,* September 2013.

## CHAPTER 2: BUILDING TRENDS

1. Tom Plant, "Thanks to Efficiency Upgrades, Home Energy Use is Back to 2001 Levels," *Advanced Energy Perspectives* (Blog), January 9, 2014.

2. U.S. Energy Information Administration, *Annual Energy Outlook 2020.*

3. Ibid.

4. Ibid.

5. Ibid.

6. Ibid.

7.  Ibid.

8.  Ibid.

9.  Ibid.

10. Ibid.

11. Eric Masanet at al, "Recalibrating global data center energy-use estimates," *Science* Volume 367 (6481):984–986, February 28, 2020.

12. Michael Belfiore, "This AI Designer Can Refine Architects' Models," *Bloomberg Businessweek,* September 21, 2017.

13. Bennett Brumson, "Construction Robots and Constructing a Robotics Community," *Robotic Industries Association* (Blog), October 11, 2007.

14. Staff Reports, "World's Fastest House Built in Montevallo," *Shelby County Reporter,* December 23, 2002.

15. "30-Story Building Built In 15 Days (Time Lapse)," Video, 2:31, January 8, 2012.

16. "Building a 15 Storey [*sic*] hotel in 6 days," Video, 4:12, November 16, 2010.

17. "Built in 48 Hours: Mohali House Enters the Record Books as the Fastest Built House of Its Kind in India," *Daily Mail Online,* December 1, 2012.

18. Tim Heffernan, "The Prefab High-Rise," *Popular Mechanics,* November 2014, 191(9): 85–86.

19. Liam Stannard, "The Robots Are Coming! Autonomous Construction Vehicles," *Big Rentz,* May, 29, 2018.

20. Carlos Balaguer and Mohamed Abderrahim, "Trends in Robotics and Automation in Construction," *Intech Open,* 2008.

21. Prashant Gopal and Heather Perlberg, "Robots May Help Build Your Next Home and Fill the Labor Gap," *Bloomberg,* April 17, 2017.

22. Mike Murphy, "You can now 3D print a house in under a day," *Quartz Media,* March 12, 2018.

23. Tamara Warren, "This cheap 3D-printed home is a start for the 1 billion who lack shelter," *The Verge*, March 12, 2018.

24. Thomas Frey, "Disposable Houses," *Futurist Speaker* (Blog), April 24, 2014.

25. Michael Murray, "Why Systems Integration Is Such a Big Problem for Smart Buildings," *Greentech Media* (Blog), October 25, 2013.

26. "Smart buildings—the future of building technology," Video, 7:26, posted by Siemens.

27. Malin Rising, "IKEA starts selling solar panels for homes," *Australian Associated Press*, October 16, 2013.

28. Rob Nikolewski, "California becomes first state requiring all new homes be built with solar," the *San Diego Union Tribune*, May 9, 2018.

29. Nicole Brown, "NYC steam system: Why we have it and how Con Ed maintains it," *AMNewYork*, July 27, 2018, https://www.amny.com/news/nyc-steam-system-1.20146953.

30. Oak Ridge National Laboratory, "Combined Heat and Power: Effective Energy Solutions for a Sustainable Future," December 1, 2008, 5.

31. Ibid.

## CHAPTER 3: SUSTAINABLE BUILDING

1. International Energy Agency, *World Energy Outlook 2012*, 327.

2. Ibid. 330.

3. Ibid. 334.

4. Ibid. 336–337.

5. Ibid. 338.

6. Ibid. 332.

7. Ibid. 331.

8. "The ZERO Code," *ZERO Code*.

9.  Katherine Ling, "Buildings Key in U.S. Climate Push Thanks to Architecture's 'Persuader in Chief'," *E&E News*, October 9, 2013: 2.

10. Ibid.

11. "Living Building Challenge," International Living Future Institute.

12. Jim Hanford et al., "Building Change," *High Performing Buildings Magazine*, Winter 2016.

13. Amory B. Lovins, *Reinventing Fire*, Hartford: Chelsea Green Publishing, 2011, 86.

14. "Nanopore Thermal Insulation," Nanopore.

15. "What is Aerogel?" *Aerogel.org* (Blog).

16. Richard Keech, "Changing phase: Are PCMs living up to their promise?" *Sanctuary Magazine*, April 16, 2018.

17. U.S. Energy Information Administration, *Annual Energy Review 2011*, Washington DC: U.S. Energy Information Administration, 2011, F2.9.

18. Sandrine Ceurstemont, "The paint that uses the sun to cool your home," *NewScientist*, October 14, 2017, 10.

19. James Temple, "A Material That Throws Heat into Space Could Soon Reinvent Air-Conditioning," *Technology Review*, September 12, 2107.

20. Justin Gerdes, "Triple-Glazed 'Super Window' Builds on Decades of Government Research," *Greentech Media*, July 2, 2018.

21. Joe Verrengia and Heather Lammers, "Smart Windows: Energy Efficiency with a View," *NREL News Feature*, January 22, 2010.

22. Kevin Bosner, "How Smart Windows Work," *HowStuffWorks*, March 29, 2001.

23. Brian Clark Howard, "Smart Glass: Windows that Lighten and Darken on Cue," *Popular Science*, January 28, 2014: 22.

24. Cate Morgan-Harlow, "3 Types of Energy Smart Windows," *BuildDirect Blog: Life At Home* (Blog), October 10, 2016.

25. Ibid.

26. Andy Extance, "The Dawn of Solar Windows," *IEEE Spectrum*, January 24, 2018.

27. Ibid.

28. "4 Things to Know About Mass Timber," *Think Wood*, April 25, 2018.

29. "What is PEFC?," Programme for the Endorsement of Forest Certification.

30. Hallie Busta, "Mass timber 101: Understanding the emerging building type," *Construction Dive*, May 24, 2017.

31. Andy Ridgway, "Skutterudites: The heat scavengers," *NewScientist*, October 11, 2014.

32. Annie Sneed, "Wood is the New Steel: Tall Timber Buildings Could Reduce Emissions," *Scientific American*, September 2017, 19.

33. Henry Fountain, "Towers of Steel? Look Again," the *New York Times*, September 24, 2013, D6, ScienceTimes.

34. Busta, "Mass timber 101."

35. Patricia Layton, "Mass timber comes of age: Code consideration, evolving supply chain promise new options for tall wood buildings," *BDC Network*, July 2, 2018, https://www.bdcnetwork.com/mass-timber-comes-age-code-consideration-evolving-supply-chain-promise-new-options-tall-wood.

36. "4 Things to Know," *Think Wood*.

37. National Renewable Energy Laboratory, "DEVap: Evolution of a New Concept in Ultra Efficient Air Conditioning," January 2011, doi: 10.2172/1004010.

38. U.S. Department of Energy, "How Energy-Efficient Light Bulbs Compare with Traditional Incandescents,".

39. Neal Elliott, Maggie Molina, and Dan Trombley, "A Defining Framework for Intelligent Efficiency," *ACEEE*, June 5, 2012.

40. U.S. Department of Energy, *A Common Definition for Zero Energy Buildings,* Oak Ridge: U.S. Department of Energy, 2015.

41. Ibid.

42. International Energy Agency, "Sustainable Development Scenario: A cleaner and more inclusive energy future," 2019.

## CHAPTER 4: SENTIENT-APPEARING BUILDINGS

1. David Rose, *Enchanted Objects: Design, Human Desire, and the Internet of Things,* New York: Simon & Schuster, 2014.

2. Cory Doctorow and Charles Stross, *The Rapture of the Nerds,* New York: Tom Doherty Associates, LLC, 2012, 12–13.

3. George Dvorsky, "Why 'Utility Fogs' Could Be the Technology That Changes the World," *IO9* (Blog), August 8, 2012.

4. Christopher Jobson, "A Self-Folding Origami Robot That Can Walk, Climb, Dig, Carry, Swim and Dissolve into Nothing," *Colossal,* June 1, 2015.

5. Kelly Weinersmith and Zach Weinersmith, *Soonish: Ten Emerging Technologies That'll Improve and/or Ruin Everything,* New York: Penguin, 2017.

6. Technische Universitaet Muenchen, "With Carbon Nanotubes, a Path to Flexible, Low Cost Sensors: Potential Applications Range from Air-Quality Monitors to Electronic Skin," *Science Daily* (Blog), September 25, 2013.

7. University Saarland, "Printed Electronics: A Multi-Touch Sensor Customizable with Scissors," *Science Daily* (Blog), October 8, 2013.

8. Laura Bliss, "The New Alchemy: How Self-Healing Materials Could Change the World," *CityLab,* September 15, 2014.

9. Bob Moffit, "Composite Building Materials for Green Building," *Ashland Inc.* (Report), March 2012.

10. Tsvi Bisk, "Unlimiting Energy's Growth," the *Futurist Magazine* 46, no. 3 (May–June 2013): 29–31.

11. Jamie Carter, "40 Ways Graphene Is About to Change Your Life" *Tech Radar,* March 13, 2019.

12. Mike Williams, "Carbon's New Champion: Carbyne, a Simple Chain of Carbon Atoms, Strongest Material of All?" *Science Daily* (Blog), October 9, 2013.

13. William McDonough and Michael Braungart, *The Upcycle: Beyond Sustainability—Designing for Abundance,* New York: North Point Press, 2013, 15.

14. Stacey Higginbotham, "The Frugal Factory," *IEEE Spectrum Magazine,* 1055, no. 10, (October 2018): 22.

15. Irene Kim and Exa Zim, "Four MIT Graduates Created a Restaurant with a Robotic Kitchen That Cooks Your Food in Three Minutes or Less," *Business Insider,* May 28.

16. "Toilet of the Future? You Might Want to Sit down for This," *NBC News,* December 7, 2013.

17. "5 Medical Robots Making a Difference in Healthcare," *Case Western Reserve University* (Blog), December 28, 2019.

18. Brian Clark Howard, "11 Vertical Farms to Transform Our Cities," *Esquire* (Blog), June 7, 2010.

19. Michaeleen Doucleff, "Vertical 'Pinkhouses:' The Future of Urban Farming?" *NPR* (Blog), May 21, 2013.

20. Ibid.

21. Ibid.

22. Ibid.

23. Ryan Nakashima, "People, Power Keep Indoor Farming Down to Earth," *Austin American-Statesman,* May 12, 2018.

24. Ibid.

25. Stefano Boeri Architetti, "Vertical Forest."

26. Stefano Boeri Architetti, "La Forêt Blanche et la Cour Verte."

27. Stefano Boeri Architetti, "Trudo Vertical Forest."

28. Stefano Boeri Architetti, "Nanjing Vertical Forest."

29. "Supertree Grove (Supertree Observatory & OCBC Skyway)," Gardens by the Bay.

30. Jane Margolies, "Making a Day at the Office Like a Walk in the Park," the *New York Times,* January 16, 2019, B5.

31. Cameron McWhirter, "Cities Covet 'Deck Parks' To Aid Cores," the *Wall Street Journal,* January 2, 2019, A3.

32. Dale Sweetnam and April Lee, "NFL Stadiums Produce Onsite Energy with Solar PV Projects," *EIA: Today in Energy,* January 31, 2014.

33. Ibid.

34. "Glossary of terms and acronyms," *California Independent System Operator.*

35. Kiley Kroh, "Introducing The First-Ever World Cup Stadium Powered By Solar," *Climate Progress,* June 14, 2014.

36. Sebastian Jordana, "Taiwan Solar Powered Stadium/Toyo Ito," *Arch Daily,* March 17, 2013.

37. Adam Vaughan, "Japan begins work on 'world's largest' floating solar farm," the *Guardian,* January 17, 2016.

38. Jason Daley, "China Turns on the World's Largest Floating Solar Farm," the *Smithsonian Magazine* (June 2017).

39. The Solar Foundation, "Brighter Future: A Study on Solar in U.S. Schools Report," *Solar Energy Industries Association* (Blog), September 18, 2014.

40. Roddy Scheer and Doug Moss, "Sun Roof: Solar Panel Shingles Come Down in Price, Gain in Popularity," *Scientific American* (Blog), April 2, 2013.

41. "Tesla Solar Roof," *Electrek* (Guide).

42. Sara Matasci, "The Tesla roof: when will it be available and should you wait for it?" *EnergySage*, January 2, 2019.

43. "The rocks here in Oman are special, this scientist says," *New York Times*, April 26, 2018.

44. Marina Miceli, "Microalgae Prove Ideal for Green Facades," *Arup* (Blog), September 14, 2012.

45. Kristin Houser, "New Artificial Leaf Design Could Absorb Far More $CO_2$," *Futurism: The Byte*, February 13, 2019.

46. Popular Science Staff, "Top green innovations of 2014: Plastic from thin air," *Popular Science*, December 1, 2014.

47. Tony Chan, "Envision 2050: The Future of Cities: Two Paths," *Ensia*, June 16, 2014.

48. Shawn Gehle, "Envision 2050: The Future of Cities: Hope for the Future," *Ensia*, June 16, 2014.

49. Josef Hargrave, "It's Alive! Can You Imagine the Urban Building of the Future?" *ARUP Foresight*, January 2013.

50. Ibid.

51. Elizabeth Dwoskin, "They're Tracking You When You Turn Off the Lights," *The Wall Street Journal*, October 20, 2014.

## CHAPTER 5: GETTING OFF OIL

1. Elizabeth Kolbert, "Hosed," *The New Yorker*, November 8, 2009.

2. Kurt Vonnegut, "Confetti #52," 2006, Kurt Vonnegut Prints.

3. Javier Blas, "Remember Peak Oil? Demand May Top Out Before Supply Does," *Bloomberg Businessweek*, July 17, 2017, 32–33.

4. Michael E. Webber, *Power Trip*, Basic Books, 2019, 109.

5. Keith Naughton and David Welch, "This Is What Peak Car Looks Like," Hyperdrive (Blog), *Bloomberg Businessweek*, February 28, 2019.

6. Andrew J. Hawkins, "Why Congestion Pricing Can Save Cities from Their Worst Possible Future," *The Verge*, March 29, 2019.

7. Naughton and Welch, 2019.

8. "Global EV Outlook 2018,"*International Energy Agency*, 11.

9. Ibid. 16.

10. Stacy C. Davis and Robert G. Boundy, "Transportation Energy Data Book: Edition 37.1," April 2019, Oak Ridge, Tennessee: Oak Ridge National Laboratory, Table 2.8.

11. U.S. Department of Energy, "All-Electric Vehicles," Washington DC: Office of Energy Efficiency and Renewable Energy.

12. Daniel Sperling. "The Case for Electric Vehicles." Scientific American 275, no. 5 (1996): 54–59.

13. J.P. Morgan, "Driving into 2025: The Future of Electric Vehicles," 2018.

14. Department of Energy, "eGallon," July 13, 2019, Washington, DC.

15. Leslie Shaffer, "Electric vehicles will soon be cheaper than regular cars because maintenance costs are lower, says Tony Seba," CNBC, June 14, 2016.

16. U.S. Department of Energy Alternative Fuels Data Center, "Alternative Fueling Stations Counts by State," Washington, DC: Office of Energy Efficiency & Renewable Energy.

17. Rachael Nealer, David Reichmuth, and Don Anair, "Cleaner Cars from Cradle to Grave," November 2015, Cambridge, MA: Union of Concerned Scientists.

18. Mike Ives, "Boom in Mining Rare Earths Poses Mounting Toxic Risks," Yale Environment 360, January 28, 2013.

19. John W. Brennan and Timothy E. Barder, "Battery Electric Vehicles vs. Internal Combustion Engines," November 2016, Boston, MA: Arthur D. Little.

20. Tyler Hamilton, "Lithium Battery Recycling Gets a Boost," MIT Technology Review, August 12, 2009, https://www.technology review.com/s/414707/lithium-battery-recycling-gets-a-boost/. Also

see Kyle Field, "Yes, Tesla Recycles All of Its Spent Batteries & Wants To Do More In the Future," Clean Technica, June 7, 2018.

21. David Stringer and Jie Ma, "Where 3 Million Electric Vehicle Batteries Will Go When They Retire," Hyperdrive (Blog), *Bloomberg Businessweek,* June 27, 2018.

22. Mitch Jacoby, "It's Time to Get Serious About Recycling Lithium-Ion Batteries," Chemical & Engineering News, July 14, 2019.

23. Mark Chediak, "Electric Buses Will Take Over Half the World Fleet by 2025," Hyperdrive (blog), *Bloomberg Businessweek,* February 1, 2018.

24. International Energy Agency, *Energy Technology Perspectives 2018,* Paris: OECD Publishing, 2018, 10.

25. Isabella Burch and Jock Gilchrist, "Survey of Global Activity to Phase Out Internal Combustion Engine Vehicles," Santa Rosa, CA: Center for Climate Protection, 2018.

26. Jon Walker, "The Self-Driving Car Timeline—Predictions from the Top 11 Global Automakers," Emerj, May 14, 2019.

27. Andrew J. Hawkins, "Riding in Waymo One, the Google Spinoff's First Self-Driving Taxi Service," *The Verge,* December 5, 2018.

28. Josh Lowensohn, "Uber Gutted Carnegie Mellon's Top Robotics Lab to Build Self-Driving Cars," *The Verge,* May 19, 2015.

29. "Autonomous Vehicles: Uncertainties and Energy Implications," Department of Energy, Washington, DC: U.S. Energy Information Administration, May 2018, 3.

30. Elizabeth Woyke, "The Blind Community Has High Hopes for Self-Driving Cars," MIT Technology Review, October 12, 2016.

31. Charlie Johnson and Jonathan Walker, "Peak Car Ownership Report," Basalt, CO: Rocky Mountain Institute, 2016.

32. Russell Brandom, "Self-Driving Cars Are Headed Toward an AI Roadblock," *The Verge,* July 3, 2018.

33. Brad Plumer, "5 Big Challenges That Self-Driving Cars Still Have to Overcome," *Vox,* April 21, 2016.

34. Aarian Marshall, "A Not-So-Sexy Plan to Win at Self-Driving Cars," *Wired,* December 4, 2018.

35. Christine Fisher, "Volvo Trucks' Autonomous Vehicle Is Hauling Goods in Sweden," Engadget, June 14, 2019.

36. Alex Davies, "The War to Remotely Control Self-Driving Cars Heats Up," *Wired,* March 26, 2019.

37. Ibid.

38. James Ayre, "Daimler Trucks Has Begun Testing Truck Platooning Tech in Japan," Clean Technica, January 31, 2018.

39. Samuel Gibbs, "Tesla Seeking to Test Driver-Free Electric Trucks on Public Roads," *The Guardian,* August 10, 2017.

40. Gillian Tett, "US truck driver shortage points to bigger problems," *Financial Times,* April 8, 2018.

41. Alan Adler, "Autonomous Guided Big Rig Platooning Could Arrive Before Robo-Cars," Trucks, June 18, 2018.

42. Brian Palmer, "Let's Make an Effort to Move More Freight by Rail and Less by Road," *Washington Post,* March 3, 2014.

43. U.S. Department of Energy, "How Much Ethanol Is In Gasoline, and How Does It Affect Fuel Economy," Frequently Asked Questions, Washington, D.C.: US Energy Information Administration, May 14, 2019.

44. Timothy Searchinger et al., "Use of U.S. Croplands for Biofuels Increases Greenhouse Gases through Emissions from Land-Use Change," *Science* 319, no. 5867 (2008): 1238–240.

45. Robert Rapier, "Cellulosic Ethanol Falling Far Short of Hype," *Forbes,* February 11, 2018.

46. European Environment Agency, "Final Energy Consumption by Mode of Transport," Copenhagen, November 2018, Fig. 4.

47. U.S. Department of Agriculture, "Brazil Biofuels Annual," Washington, DC: Global Agriculture Information Network, October 2018.

48. University of Utah, "Turning Algae into Fuel," EurekAlert (Press Release), AAAS.

49. Amy Nordrum, "New Tech Could Turn Algae Into the Climate's Slimy Savior," IEEE Spectrum, May 30, 2018.

50. Lisa Baertlein, "UPS to Spend $130 Mln on New Natural Gas Vehicles, Fueling Stations," *Reuters,* June 19, 2018.

51. Gerald Porter Jr., "The Race to Fuel the Buses of Future Is On," Hyperdrive (blog), *Bloomberg,* July 8, 2019.

52. Ford Motor Company, "Presenting Ford New Aspire CNG."

53. Steven Overly, "Seven Automotive Trends to Watch in 2017," *The Washington Post,* January 14, 2017.

54. U.S. Energy Information Administration, *Annual Energy Outlook 2014,* Washington DC: U.S. Energy Information Administration, 2014.

55. Ibid.

56. Vaclav Smil, *Prime Movers of Globalization,* Cambridge, MA: MIT Press, 2010.

57. Andrew Hawkins, "Electric Flight Is Coming, but the Batteries Aren't Ready," *The Verge,* August 14, 2018.

58. Hazel Plush, "Revealed: What airlines really spend your money on," *The Telegraph,* May 23, 2016.

59. IATA, "IATA Forecasts Passenger Demand to Double Over 20 Years," 2016.

60. FAA, "FAA Aerospace Forecast Fiscal Years 2017–2037," 2017.

61. Carolyn Davidson et al., "An Overview of Aviation Fuel Markets for Biofuels Stakeholders," July 2014.

62. Tom Ravenscroft, "New Road in Sweden Charges Electric Cars As They Drive Along," *Dezeen*, April 23, 2018.

63. Debbie Hammel, "Raising the Bar: NRDC's 2017 Aviation Biofuels Scorecard," The Natural Resources Defense Council, October 24, 2017.

64. IATA, "Fact Sheet Climate Change & CORSIA," 2017.

65. Nancy N. Young, "The Future of Biofuels: U.S. (and Global) Airlines," Aviation Alternative Fuels, 2014 EIA Conference

66. Emily Newes, Jeongwoo Han, and Steve Peterson, "Potential Avenues for Significant Biofuels Penetration in the U.S. Aviation Market," 2017.

67. IATA, "IATA Sustainable Aviation Fuel Roadmap," 2015.

68. John Fialka, "Are Advanced Biofuels for Airplanes Ready for Takeoff?" *Scientific American*, May 22, 2017.

69. Fred Lambert, "All-Electric Ferry Cuts Emission By 95% and Costs By 80%, Brings in 53 Additional Orders," *Electrek*, February 3, 2018.

70. Red and White Fleet, "Enhydra,".

71. Starre Vartan, "The Future of Seattle's Ferries Is Electric," Our Stories (blog), Natural Resources Defense Council, December 26, 2018.

72. Vaclav Smil, "Electric Container Ships Are a Hard Sail," IEEE Spectrum, February 25, 2019.

73. Captain Maritime and Offshore News

74. Steve Hanley, "Eco Marine Power Testing Solar Sails for Ocean Going Cargo Ships," Clean Technica, January 31, 2018.

75. Maria Gallucci, "Cutting Carbon in Maritime Cargo Shipping," Texas Energy Symposium, October 19, 2017.

## CHAPTER 6: SENTIENT-APPEARING TRANSPORTATION

1. Glen Larson, *Knight Rider,* Glenn Larson Productions, 1982–1986, NBC.

2. Michael Bay, *Transformers,* DreamWorks Pictures, 2007 and William Hanna and Joseph Barbera, *The Jetsons,* Hanna-Barbera Productions, 1962–1987, ABC.

3. Charles McLellan, "What Is V2X Communication? Creating Connectivity for the Autonomous Car Era," *ZDNet,* March 12, 2018.

4. Adam Fisher, "Google's Self-Driving Cars: A Quest for Acceptance," *Popular Science,* September 18, 2013.

5. Loes Witschge, "Rotterdam Is Building the Most Automated Port in the World," *Wired,* October 7, 2019.

6. Jennifer Jones, "10 Largest Drones in the World," Largest.org, September 5, 2019.

7. Dan Maloney, "Automate the Freight: Autonomous Delivery Hits the Mainstream," Hackaday, June 17, 2019.

8. Dain Evans, "Cargo Drones Could Be the Future of the Shipping Industry," *CNBC,* July 29, 2018.

9. Lara Kolodny and Darren Weaver, "These Drones Can Haul a 20-Pound Load for 500 Miles and Land on a Moving Target," *CNBC,* May 26, 2018.

10. Bernard Marr, "Rolls-Royce And Google Partner To Create Smarter, Autonomous Ships Based On AI And Machine Learning," *Forbes,* October 23, 2017.

11. Oskar Levander, "Forget Autonomous Cars—Autonomous Ships Are Almost Here," *IEEE Spectrum,* January 28, 2017.

12. Michelle Yan, "Watch Ford's Delivery Robot That Walks On Two Legs Like a Human," *Business Insider,* May 23, 2019.

13. Ken Washington, "Meet Digit: A Smart Little Robot That Could Change the Way Self-Driving Cars Make Deliveries," *Medium,* May 22, 2019.

14. Brad Templeton, "How Might Self-Driving Cars Refuel Themselves?" *Quora*, June 10, 2015.

15. Joe Pappalardo, "The Unmanned X-47B Just Pulled Off a Mid-Air Refueling," *Popular Mechanics*, April 22, 2015.

16. Brian Metzger, "Remote Autonomous Refueling Buoy," *TechLink*.

17. Jonas Thor Olsen, "A Neste Petrol Station in Finland Uses a Robot ARM to Fuel People's Cars," *Petrol Plaza*, June 18, 2019.

18. Tracey Lien, "Uber Says It Will Bring Its Flying Taxis to Los Angeles in 2020," *Los Angeles Times*, November 8, 2017.

19. Bruce Gellerman, "It's Electri-Flying: Cape Air Pioneers Flights Without Fossil Fuels," WBUR, August 8, 2019.

20. Andrew Hawkins, "Watch This All-Electric 'Flying-Car' Take Its First Flight in Germany," *The Verge*, April 20, 2017.

21. Robert Wall, "Electric Planes Gain Momentum," *The Wall Street Journal*, June 17, 2016.

22. Mike Cherney, "Hybrid Planes Take Electric Flight," *The Wall Street Journal*, July 15, 2019.

23. Tim Bowler, "Why the Age of Flight Is Finally Upon Us," *BBC News*, July 3, 2019.

24. FAA, "Fact Sheet—Supersonic Flight," December 18, 2019.

25. Jacopo Prisco, "Concordski: What ever happened to Soviets' spectacular rival to Concorde?" CNN, July 2019.

26. David Slotnick, "The Concorde made its final flight 16 years ago and supersonic air travel has yet to return—here's a look at its awesome history," *Business Insider*, October 7, 2019.

27. Jeremy Bogaisky, "Boom Raises $100M To Develop A Supersonic Airliner. It's Going To Need A Whole Lot More," *Forbes*, January 4, 2019.

28. Tom Chitty, "In Pictures: Here Are the Planes Being Built to Bring Back Supersonic Travel," *CNBC*, January 18, 2019.

29. Alexandra Ossala, "Are Luxury Travelers Ready to Shell Out For Shorter Supersonic Flights," *Quartz,* March 28, 2019.

30. David Hoerr, "Point-to-point Suborbital Transportation: Sounds Good On Paper But..." *The Space Review,* May 5, 2008.

31. Alex Davies, "Elon Musk's Rocket Travel Plan Is Def Possible, Def Bananas," *Wired,* September 29, 2017.

32. Jeff Foust, "Blue Origin Plans to Start Selling Tickets in 2019 for Suborbital Spaceflights," *Space News,* July 10, 2018.

33. Robert A. Heinlin, "Friday, 1982," Phoenix Pick Edition, November, 2017.

34. Jeremy White, "Forget Supersonic, the Future of Super-Fast Flight Is Sub-Orbital," *Wired,* September 30, 2018.

35. BBC Radio, "Elon Musk's Hyperloop and Brunel's Atmospheric Traction Rail," (Video), January 29, 2019.

36. Robert M. Salter, "The Very High Speed Transit System," RAND Corporation, August 1972.

37. SpaceX, "The Official SpaceX Hyperloop Pod Competition," .

38. Ruqayyah Moynihan and Amira Ehrhardt, "A Dutch Startup Has Made Europe's First Functioning Hyperloop System, Six Years After Elon Musk Hinted That He Wanted to Do the Same," *Business Insider,* July 30, 2019.

39. CNN, "How Long Until Hyperloop Is Here?" December 10, 2019.

40. Andrew Hawkins, "Hyperloop Project in India Inches Closer to Reality," *The Verge,* July 21, 2019.

41. Ed Blazina, "State to Begin Study of Hyperloop Technology, Potential Pittsburgh-to-Philadelphia Route," *Pittsburgh Post-Gazette,* March 9, 2019.

42. Edward McKinley, "Hyperloop One Takes First Step on a Promising Test Road in America's Heartland," CNBC, October 18, 2018.

43. Blazina, 2019.

44. Peter Farquhar, "Richard Branson Just Unveiled His Vision for Virgin Hyperloop One—And It Looks Straight Out of a Sci-Fi Film," *Insider*, May 1, 2018, https://www.insider.com/richard-branson-just-unveiled-his-vision-for-virgin-hyperloop-2018-5.

45. Victoria Burnett, "Near Mexico City, Cable Car Lets Commuters Glide Over Traffic," *New York Times*, December 28, 2016

46. "'Flying Whale' Blimp That Never Lands Joins Global Cargo Airship Race," *Transport Topics*, March 27, 2018.

47. Eric Adams, "Lockheed's Hybrid *Airship* Is Part Blimp, Part Hovercraft, No Hot Air," *Wired*, October 7, 2016.

48. Craig Neal, "The Emergence of Cargo Airships: An Opportunity for Airports," *International Airport Review*, July 18, 2017.

49. Ashlee Vance, "These Giant Printers Are Meant to Make Rockets," *Bloomberg Businessweek*, October 18, 2017.

50. Kevin Bonsor, "How Space Elevators Will Work," *How Stuff Works*, October 6, 2000.

51. Kelly Weinersmith, Zach Weinersmith, *Soonish: Ten Emerging Technologies That'll Improve And/Or Ruin Everything*, Penguin, 2017, 35.

52. Alison King, "ThothX Releases Space Elevator Animation" (Press Release), *Thoth Technology*, August 22, 2016.

53. Charles H. Bennett et al., "Teleporting an Unknown Quantum State via Dual Classical and Einstein-Podolsky-Rosen Channels," *Physical Review Letters* 70, no. 13 (1993): 1895.

54. Kevin Bonsor and Robert Lamb, "How Teleportation Will Work," *How Stuff Works*, October 25, 2000.

55. Alex Knapp, "Physicists Quantum Teleport Photons Over 88 Miles," *Forbes* (Blog), September 6, 2012.

56. Ibid.

57. Kathryn Doyle, "Why Don't We Have Teleportation?" *Popular Mechanics*, January 24, 2013.

58. Ibid.

59. Ibid.

60. Michio Kaku, *The Future of the Mind: The Scientific Quest to Understand, Enhance, and Empower the Mind,* Anchor Books, 2015, 285.

61. Kaku, 290.

## CHAPTER 7: THE CHANGING POWER INDUSTRY

1.  U.S. Energy Information Administration, *Annual Energy Outlook 2019.*

2.  International Energy Agency, *World Economic Outlook, 2016,* 2017, 63.

3.  Brian Park, "In 2015, U.S. Coal Production, Consumption, and Employment Fell by More Than 10%," *EIA: Today in Energy,* November 10, 2016.

4.  Christine Chan and Timothy Gardner, "Coal Outlook Bleak," *Reuters Graphics,* November 13, 2017.

5.  U.S. Energy Information Administration, 2019.

6.  Gwynne Dyer, "Japan, Germany and the War on Rationality," *The Japan Times,* Feb. 21, 2020.

7.  International Energy Agency, *World Energy Outlook, 2016,* Organization for Economic Cooperation and Development, 2017, 204

8.  Andreas Rinke and Chris Gallagher, "Germany's Merkel Signals Support for 2038 Coal Exit Deadline," *Reuters,* February 4, 2019.

9.  "Medium-Term Coal Market Report 2016," *International Energy Agency.*

10. International Energy Agency, *World Energy Outlook* 2016, Table 5.2.

11. River Davis, "Cheaper Natural Gas Tackles Coal in East Asia," *The Wall Street Journal,* April 20, 2020.

12. Jamie Condliffe, "Clean Coal's Flagship Project Has Failed," *MIT Technology Review*, June 29, 2017.

13. Matthew L. Wald, "Higher Costs Cited as US Shuts Down Coal Project," *New York Times*, January 31, 2008.

14. Global Carbon Capture and Storage Institute, "Carbon Capture and Storage: A Key Technology for a Decarbonized Future," June 10, 2019.

15. Brad Plumer, "Coal, in a First, Will Be Passed By Renewables," *New York Times*, May 14, 2020.

16. "How Much Natural Gas Does the United States Have, and How Long Will It Last?" U.S. Energy Information Administration, April 5, 2019.

17. U.S. Energy Information Administration, *Annual Energy Outlook 2019*, Washington, DC: U.S. Energy Information Administration, 2018, slide 77.

18. Ibid. Slide 73.

19. International Energy Agency, *World Energy Outlook 2017*, 52.

20. Henry Fountain, "Unlocking the Potential of 'Flammable Ice'," *New York Times*, September 16, 2013.

21. Gavin Bade, "Santee Cooper, Scana Abandon Summer Nuclear Plant Construction," *Utility Dive*, July 31, 2017.

22. Union of Concerned Scientists, "The Nuclear Power Dilemma," Oct. 9, 2018.

23. "Nuclear Power in the USA," *World Nuclear Association*, March 2020.

24. "Nuclear Costs in Context," Nuclear Energy Institute, September, 2019.

25. Agneta Rising, "World Nuclear Performance Report 2019," World Nuclear Association, 2019.

26. Alexander Hurst, "France Could Close Up to 17 Nuclear Reactors by 2025," *France 24*, October 7, 2017.

27. "Nuclear Power in China," *World Nuclear Association*, April, 2020.

28. "Nuclear Power in Russia," *World Nuclear Association*, April, 2020.

29. "Nuclear Power in the United Kingdom," *World Nuclear Association*, July 2019.

30. "Solar, Wind, and Nuclear have amazingly low Carbon Footprints", *Carbon* Brief, December 2017.

31. Andrew Meyers, "Wind Could Meet World's Total Power Demand—and Then Some—by 2030," *Stanford Engineering: Research & Ideas*, September 10, 2012.

32. American Wind Energy Association, "Wind Energy in the United States."

33. World Wind Energy Association, "Wind Power Capacity Worldwide Reaches 597 GW, 50,1 GW Added in 2018," February 25, 2019.

34. Robert Fares, "Wind Energy Is One of the Cheapest Sources of Electricity, and It's Getting Cheaper," *Plugged In* (Blog), *Scientific American*, August 28, 2017.

35. Jim Malewitz, "$7 Billion Wind Power Project Nears Finish," *The Texas Tribune*, October 14, 2013.

36. Electric Reliability Council of Texas, *Quick Facts*, January 2019.

37. Jack Money, "Oklahoma Grid Operator Says Wind Provided More Than 60% of the Power Friday," *The Oklahoman*, March 17, 2018.

38. International Renewable Energy Agency, *Renewable Capacity Statistics 2018.*

39. "Dutch Open 'World's Largest Offshore' Wind Farm," Phys.org, May 8, 2017.

40. Orsted, "Wind projects: Block Island Wind Farm."

41. Hiroko Tabbuchi, "To Expand Offshore Power, Japan Builds Floating Windmills," *New York Times*, October 24, 2013.

42. GE Renewable Energy, "Haliade-X Offshore Wind Turbine Platform."

43. National Renewable Energy Laboratory, "Banking on Solar: New Opportunities for Lending," Department of Energy, August 2014.

44. "First Solar Automates and Profits," *Premier Gazette*, January 24, 2018, https://premiergazette.com/2018/01/first-solar-automates-and-profits/.

45. International Renewable Energy Agency, *Renewable Capacity Statistics 2019*.

46. Steve Hanley, "New Solar Price Record: Tucson Utility Inks Deal for Solar Power That Costs Less Than 3 Cents per Kilowatt-Hour!" Clean Technica, May 24, 2017.

47. Cara Marcy, "Solar Photovoltaic Costs Are Declining, But Estimates Vary Across Sources," *EIA: Today in Energy*, March 21, 2018.

48. Ran Fu et al., *US Solar Photovoltaic System Cost Benchmark: Q1 2016*, Golden, CO: National Renewable Energy Lab, September 1, 2016.

49. Rob Wile, "SHELL: Solar Will Be the World's No. 1 Source of Energy By the End of the Century," *Business Insider*, October 1, 2013.

50. UNESCO World Water Assessment Programme, "Great World Hydropower Potential."

51. Neevatika Verma et al., "Solid State Transformer for Electrical System: Challenges and Solution," *2018 2nd International Conference on Electronics, Materials Engineering & Nano-Technology (IEMENTech)*, 1–5. IEEE, 2018.

52. Markets and Markets, "Smart Grid Market by Software (AMI, Grid Distribution, Grid Network, Grid Asset, Grid Security, Substation Automation, and Billing & CIS), Hardware (Smart Meter), Service (Consulting, Integration, and Support), and Region—Global Forecast to 2023," Smart Grid Market.

53. Gavin Bade, "FERC Asks Grid Operators For More Detail on Storage Participation," *Utility Dive*, April 3, 2019.

54. Frank Andorka, "New York Becomes the Fourth State to Add Energy Storage Mandate," *PV Magazine USA*, December 1, 2017.

55. Fred Lambert, "Tesla's Massive PowerPack Battery…17 million", *Electrek*, September 24, 2018.

56. FP&L Newsroom, March 25, 2019.

57. Robert Dieterich, "24-Hour Solar Energy: Molten Salt Makes It Possible," *Inside Climate News*, Jan. 16, 2018.

58. Christian Roselund, "Ice Energy brings the deep freeze to U.S. energy storage," *PV Magazine*, Feb. 13, 2019.

59. Julian Spector, "Ice Energy will launch ice storage in first quarter 2017," *Greentech Media*, Oct. 6, 2016.

60. "Key World Energy Statistics 2018," International Energy Agency, 2018.

61. "Electricity Net Generation: Total (All Sectors), 1949–2011," Annual Energy Review 2011, U.S. Energy Information Administration, 2012.

62. Jürgen Weiss et al., "Electrification: Emerging Opportunities for Utility Growth," Brattle Group, January 2017.

63. International Energy Agency, "Global EV Outlook," Paris, France, 2019.

64. Xiaodan Xu, H. M. Abdul Aziz, and Randall Guensler, "A modal-based approach for estimating electric vehicle energy consumption in transportation networks," *Transportation Research Part D: Transport and Environment*, October 2019, Vol 72, 249–264.

65. Digiconomist, "Bitcoin Energy Consumption Index."

66. Richard Chirgwin, "By 2040, Computers Will Need More Electricity Than The World Can Generate," the *Register*, July 25, 2016.

67. Digiconomist, 2019.

68. Ibid.

69. Evan Mills, "The Carbon Footprint of Indoor Cannabis Production," *Energy Policy* 46, 58–67.

70. EUCI, "Cannabis industry drives growing electricity demand, expected to soar with increased legalization," October 31, 2018.

71. Christopher Mims, "Are Shipping Containers the Future of Farming?" the *Wall Street Journal*, June 8, 2016.

72. Chris Michael, "The Good, the Bad, and the Ugly of Container Farms," *Medium* (Blog), March 16, 2017

73. Fred Beach et al., "A Comparison of New Electric Utility Business Models," LBJ School of Public Affairs, Policy Research Project Report, Number 191, Energy Institute, April, 2017.

74. Joshua Emerson Smith, "Market Transformation Will End Dominance of Electrical Utilities, Regulators Predict," the *San Diego Union-Tribune*, July 16, 2017.

## CHAPTER 8: CLEAN ENERGY SOLUTIONS

1. U.S. Energy Information Administration, *Annual Energy Outlook 2018*, Washington, DC: U.S. Energy Information Administration, 2018.

2. Renewables 2018, *Global Status Report*, Renewable Energy Policy Network for the 21st Century, 2018.

3. International Energy Agency, *Renewables Information 2017*, Paris, France, 2017.

4. BP, *BP Statistical Review of World Energy 2019*, 68th Edition

5. International Energy Agency, *Energy Technology Perspectives 2015*, Paris: OECD Publishing, 2015.

6. International Energy Agency, *Energy Technology Perspectives 2017*, Paris: OECD Publishing, 2017.

7. The World Bank, "Renewable energy Consumption (% of total final energy consumption) 1990–2015.

8.  Stephen Pacala and Robert Socolow, "Stabilization Wedges: Solving the Climate Problem for the Next 50 Years with Current Technologies," *Science* 305, No. 5686 (2004): 968–972.

9.  Ibid.

10. Mark Z. Jacobson and M. A. Delucchi, "A Worldwide Plan to Eliminate Global Warming, Air Pollution, and Energy Instability With Wind, Water, and Sunlight (WWS)," AGU Fall Meeting Abstracts, 2011.

11. Ibid.

12. Mark Z. Jacobson et al., "100% Clean and Renewable Wind, Water, and Sunlight (WWS) All-Sector Energy Roadmaps for 53 Towns and Cities in North America," *Sustainable Cities and Society* 42 (2018): 22–37.

13. C. T. Clack et al., "Evaluation of a Proposal for Reliable Low-Cost Grid Power with 100% Wind, Water, and Solar," *Proceedings of the National Academy of Sciences* 114, no. 26 (2017): 6722–6727.

14. Richard Heinberg and David Fridley, *Our Renewable Future: Laying the Path for One Hundred Percent Clean Energy*, Washington, DC: Island Press, 2016.

15. Tim Collins, "China's £115 Million Boost for Floating Nuclear Plants," *Daily Mail*, Associated Newspapers, August 21, 2017.

## CHAPTER 9: ADVANCED POWER SYSTEMS

1.  Robert B. Laughlin, *Powering the Future: How We Will (Eventually) Solve the Energy Crisis and Fuel the Civilization of Tomorrow*, New York: Basic Books, 2011, 92.

2.  Anna Possner, "Open-Ocean Wind Farms," (Press Release) *American Association for the Advancement of Science*, October 9, 2017.

3.  Orbital Marine Power, "SR2000,"

4.  Power Technology, "West Islay Tidal Farm,".

5.  Robert B. Laughlin, 2011, 113.

6.  Ibid, 104.

7.  Brian Wang, "Terrestrial Energy Notifies Nuclear Regulator of Planned 2019 Molten Salt Reactor Licensing Application," *Next Big Future,* January 25, 2017.

8.  "World's First Floating Nuclear Power Plant Bound For the Arctic, Warns Greenpeace," (Press Release) *Greenpeace International,* April 28, 2018.

9.  Paul Rincon, "Nuclear Fusion Milestone Passed At US Lab," *BBC News,* October 7, 2013.

10. Daniel Clery, "Europe Focuses Fusion Research on Building a Working Power Reactor," *Science,* (2014): 127–127.

11. Anna Hirtenstein, "Nuclear Fusion Unlikely to Challenge Solar, Wind Power," *Bloomberg Businessweek,* October 20, 2017.

12. Joseph Trevithick, "Lockheed Martin Now Has a Patent For Its Potentially World Changing Fusion Reactor," the *Drive,* March 26, 2018.

13. Thomas Hornigold, "New MIT Startup Targets Working Fusion Reactor in 15 Years. Can It Be Done?" *Singularity Hub,* March 20, 2018.

14. Eric J. Lerner et al., "Confined Ion Energy > 200 keV and Increased Fusion Yield in a DPF With Monolithic Tungsten Electrodes and Pre-ionization," *Physics of Plasmas* 24, no. 10 (2017): 102708.

15. Helion Energy, "Our Technology," https://www.helionenergy.com/technology/.

16. R. J. M. Vullers et al., "Micropower Energy Harvesting," *Solid-State Electronics* 53, no. 7 (2009): 684-693.

17. Diane Toomey, "Could Evaporation Be a Significant Source of Renewable Energy?" *Yale E360,* September 28, 2017.

18. Uncharted Play, Inc., "Sockett: The Energy-Harnessing Soccer Ball," *Kickstarter.*

19. Buckminster Fuller, "What Is the Planet's Critical Path," Global Energy Network Institute.

20. Mehran Moalem, "We Could Power the Entire World By Harnessing Solar Energy From 1% of the Sahara," *Forbes*, September 22, 2016.

## CHAPTER 10: OUR CRYSTAL BALL

1.  Joel Hruska, "Quadriplegic Man Uses Exoskeleton to Walk, Move Arms Again," *Extreme Tech*, October 4, 2019.

2.  Vaclav Smil, *Prime Movers of Globalization*, Cambridge: MIT Press, 2010.

3.  Vaclav Smil, "What I See When I See a Wind Turbine," *IEEE Spectrum* 53, no. 3 (2016): 27-27.

4.  "Peek into wooden mast reveals wind power's towering future", eenews.net, May 7, 2020.

5.  Smil, *IEEE Spectrum*.

6.  Lucy Rogers, "Climate Change: The Massive $CO_2$ Emitter You May Not Know About," *BBC News*, December 17, 2018.

7.  Ibid.

8.  Ibid.

9.  Heinberg and Fridley, 2016, 96–102.

10. L. Lemay, "Coal Combustion Products in Green Building," *In Coal Combustion Products* (CCP's), 395–414, Woodhead Publishing, 2017.

11. "Study on the Review of the List of Critical Raw Materials," European Commission, Brussels (2017).

12. Tim Treadgold, "Cobalt: The Achilles Heel for Electric Car Makers," *Forbes*, March 7, 2018.

13. Stella Soon, "As Electric Vehicle Production Ramps up Worldwide, a Supply Crunch for Battery Materials Is Looming," *CNBC*, July 26, 2019.

14. Josh Gabbatiss, "'Endangered' Elements Used to Make Mobile Phones Are Running out Quickly, Scientists Warn," *The Independent*, January 22, 2019.

# INDEX

# ABOUT THE AUTHORS

**Roger Duncan** is a former Research Fellow at the Energy Institute at the University of Texas at Austin. He is the former General Manager of Austin Energy, the municipal electric utility for Austin, Texas. Prior to that, he served as executive manager for several City of Austin departments, including the Environmental and Conservation Services department and Planning and Transportation. Roger was also elected to two terms as Austin City Council member in the early 1980's. In 2005, *Business*  *Week* magazine recognized Roger as one of the 20 leading "carbon reducers" in the world, and in 2009 *National Geographic* recognized him as an international thought leader in energy efficiency.

**Dr. Michael E. Webber** serves as the Chief Science and Technology Officer at ENGIE, a global energy & infrastructure services company. Webber is also the Josey Centennial Professor in Energy Resources at the University of Texas at Austin. Webber's expertise spans research and education at the convergence of engineering, policy, and commercialization on topics related to innovation, energy, and the environment. His latest book, *Power Trip: the Story of Energy,* was published in 2019 by  Basic Books with a **6-part companion series on PBS.** His first book, *Thirst for Power: Energy, Water and Human Survival,* which addresses the connection between earth's most valuable resources and offers a hopeful approach toward a sustainable future, was published in 2016 by Yale Press and was converted into a documentary. He was selected as a Fellow of American Society of Mechanical Engineers and as a member of the 4th class of the Presidential Leadership Scholars, a leadership training program organized by Presidents George W. Bush and William J. Clinton. Webber has authored more than 400 publications, holds 6 patents, and serves on the advisory board for *Scientific American*. Webber holds a B.S. and B.A. from UT Austin, and M.S. and Ph.D. in mechanical engineering from Stanford University.

CPSIA information can be obtained
at www.ICGtesting.com
Printed in the USA
BVHW072113310820
587605BV00002B/8/J